The Fragile Balance: Understanding Earth and Environmental Science

The Fragile Balance: Understanding Earth and Environmental Science

Kenneth Caraballo

Published by Kenneth Caraballo, 2023.

THE FRAGILE BALANCE: UNDERSTANDING EARTH AND ENVIRONMENTAL SCIENCE

First edition. May 11, 2023.

Written by Kenneth Caraballo.

Table of Contents

Synopsis: In "The Fragile Balance," readers will be taken on a journey to explore the complexities of our planet's delicate balance and how it is being affected by human activities. The book will delve into various fields of Earth and Environmental Science, including geology, meteorology, oceanography, and ecology.

The book will start with an introduction to Earth's history and how it has shaped the planet's geology, climate, and ecosystems. The readers will learn about the Earth's natural systems and how they interact to create a balance that sustains life on our planet.

Next, the book will explore how humans have impacted the Earth's environment, including topics like climate change, pollution, deforestation, and overfishing. The readers will learn about the science behind these issues, their causes and consequences, and the potential solutions to address them.

The book will also cover how humans can work towards creating a sustainable future, highlighting technologies and strategies that can help us reduce our impact on the environment. It will include sections on renewable energy, green technology, sustainable agriculture, and conservation efforts.

Throughout the book, readers will encounter real-world examples and case studies that illustrate the concepts discussed. There will also be interviews with scientists and experts in the field, providing their perspectives and insights on the current state of the Earth's environment and what can be done to protect it.

"The Fragile Balance" aims to provide readers with a comprehensive understanding of Earth and Environmental

Science and how it relates to our everyday lives. The book will inspire readers to take action and make changes in their own lives to protect the environment and ensure a sustainable future for generations to come.

Chapter 1: Introduction to Earth and Environmental Science

Earth and Environmental Science is a field that studies the Earth and its surroundings, including its physical, chemical, and biological systems, and the interactions between them. This field encompasses many areas of study, including geology, meteorology, oceanography, ecology, and many more.

The study of Earth and Environmental Science is essential in understanding the complex systems that make up our planet, how they function, and how they are being impacted by human activities. As we continue to face environmental challenges such as climate change, pollution, and deforestation, the need to understand these systems and find ways to protect them becomes increasingly important.

In this chapter, we will explore the basic concepts and principles of Earth and Environmental Science and how they relate to the planet we call home.

The Formation of the Earth The Earth is estimated to be 4.5 billion years old. Scientists believe that it formed from a cloud of gas and dust left over after the formation of the sun. Over time, the particles in this cloud began to come together, forming larger and larger bodies, eventually leading to the formation of the Earth.

The Earth's Atmosphere

The Earth's atmosphere is composed of several layers, each with its own unique properties. The troposphere, the layer closest to the Earth's surface, is where weather occurs and where most of the Earth's air mass is found. The stratosphere, located above the troposphere, contains the ozone layer, which helps to protect the Earth from the sun's harmful UV radiation. The mesosphere, thermosphere, and exosphere are the other layers that make up the Earth's atmosphere.

The Earth's Oceans

The Earth's oceans cover approximately 71% of the planet's surface and contain over 97% of the Earth's water. The oceans play a crucial role in regulating the Earth's climate, absorbing heat and carbon dioxide from the atmosphere. The oceans also support a vast array of marine life and provide important resources such as food and minerals.

The Earth's Geology

Geology is the study of the Earth's physical structure and materials, including rocks, minerals, and the processes that shape them. The Earth's geology is complex and constantly changing, with processes such as plate tectonics, volcanic eruptions, and erosion shaping the planet's surface.

The Earth's Climate

The Earth's climate is determined by a variety of factors, including the amount of energy received from the sun, the composition of the atmosphere, and the Earth's orbit and rotation. The Earth's climate has fluctuated over time, with periods of warmth and cooling, but human activities are currently causing rapid changes in the climate that are leading to environmental problems such as rising sea levels and extreme weather events.

The Earth's Ecosystems

An ecosystem is a community of living organisms and their physical environment. The Earth has many different ecosystems, ranging from deserts to rainforests, each with its own unique characteristics and biodiversity. These ecosystems provide essential services such as air and water purification, nutrient cycling, and carbon storage, which are vital to human survival.

In conclusion, Earth and Environmental Science is a vast and complex field that encompasses many different areas of study. Understanding the systems that make up our planet and how they are being impacted by human activities is essential in finding ways to protect our environment and ensure a sustainable future for generations to come.

Chapter 2: The Formation of the Earth

The Earth is a complex and fascinating planet, and its origins can be traced back to the early days of the solar system. The formation of the Earth is a process that occurred over millions of years, and scientists continue to study this process in order to better understand the planet and its place in the universe.

The Nebular Hypothesis

The most widely accepted theory about the formation of the solar system is the nebular hypothesis, which states that the sun and planets formed from a rotating cloud of gas and dust called the solar nebula. According to this theory, the solar nebula was made up of 98% hydrogen and helium, with smaller amounts of other elements such as carbon, nitrogen, and oxygen.

As the solar nebula rotated, it began to flatten into a disk shape, with the sun forming at the center. Over time, the material in the disk began to coalesce into larger and larger bodies, eventually leading to the formation of the planets.

The Formation of the Earth

The formation of the Earth is believed to have occurred approximately 4.5 billion years ago. Scientists believe that the Earth formed from the same cloud of gas and dust that gave rise to the sun and other planets.

As the solar nebula began to collapse, it formed a dense central region that eventually became the sun. The remaining material in the disk began to coalesce into small particles, which collided and stuck together to form larger bodies called planetesimals. These planetesimals continued to grow through a process called accretion, in which they collided and merged to form even larger bodies.

The Earth is believed to have formed from the collision and merger of many of these planetesimals. Over time, the Earth grew in size and its gravity began to pull in other objects from the surrounding space, such as asteroids and comets.

Differentiation and the Formation of the Earth's Layers As the Earth grew in size, its interior began to heat up due to the energy released by the collisions that were taking place. This heat caused the interior of the Earth to melt, and the denser materials sink towards the center, while the lighter materials rose towards the surface. This process is known as differentiation, and it led to the formation of the Earth's layers.

The Earth's interior is made up of three main layers: the crust, the mantle, and the core. The crust is the outermost layer and is made up of a combination of solid and molten rock. The mantle is the middle layer and is mostly composed of solid rock. The core is the innermost layer and is made up of a combination of liquid and solid metal.

Conclusion

The formation of the Earth is a complex and fascinating process that occurred over millions of years. The Earth's origins can be traced back to the early days of the solar system, and its formation is closely tied to the formation of the sun and other planets. Understanding the process of Earth's formation is important not only for gaining a better understanding of our planet, but also for understanding the origins of life on Earth and the possibilities for life on other planets.

Chapter 3: The Earth's Atmosphere

The Earth's atmosphere is a vital component of our planet, and it plays a crucial role in regulating our climate and supporting life. In this chapter, we will explore the composition and structure of the Earth's atmosphere, as well as its role in protecting our planet from harmful solar radiation.

Composition of the Atmosphere

The Earth's atmosphere is composed of a mixture of gasses, with the most abundant gasses being nitrogen (78%) and oxygen (21%). Other gasses in the atmosphere include carbon dioxide (0.04%), water vapor, and trace amounts of other gasses such as methane, helium, and neon.

The atmosphere also contains small particles such as dust, pollen, and salt, which can have an impact on air quality and climate.

Layers of the Atmosphere

The Earth's atmosphere is divided into several layers based on temperature changes with altitude. The lowest layer, closest to the Earth's surface, is the troposphere. This is where weather occurs, and it contains 80% of the mass of the atmosphere.

Above the troposphere is the stratosphere, which contains the ozone layer. The ozone layer plays a crucial role in absorbing harmful ultraviolet radiation from the sun.

Beyond the stratosphere is the mesosphere, which is the coldest layer of the atmosphere. Above the mesosphere is the thermosphere, which is the hottest layer of the atmosphere and where the auroras occur.

The Exosphere is the outermost layer of the atmosphere, where the Earth's atmosphere gradually fades into space.

Role of the Atmosphere

The Earth's atmosphere plays several vital roles in regulating our planet's climate and supporting life. Some of the key functions of the atmosphere include:

1. **Regulating Temperature:** The atmosphere helps to regulate the temperature of the Earth by absorbing and trapping solar radiation.
2. **Protecting from Harmful Radiation:** The ozone layer in the stratosphere absorbs harmful ultraviolet radiation from the sun, protecting life on Earth from its harmful effects.
3. **Supporting Life:** The atmosphere provides oxygen for respiration and protects living organisms from harmful solar radiation.
4. **Maintaining Water Cycle:** The atmosphere plays a vital role in the water cycle by transporting and redistributing water vapor around the planet.
5. **Influencing Climate:** The composition and properties of the atmosphere have a significant impact on the Earth's climate and weather patterns.

Human Impacts on the Atmosphere

Human activities have had a significant impact on the Earth's atmosphere. The burning of fossil fuels and other human activities have released large amounts of carbon dioxide into the atmosphere, leading to global warming and climate change.

Human activities have also caused damage to the ozone layer, particularly through the use of chlorofluorocarbons (CFCs) in refrigerants and other products. The Montreal Protocol, an international agreement signed in 1987, has helped to reduce the use of CFCs and protect the ozone layer.

Conclusion

The Earth's atmosphere is a complex and dynamic system that plays a crucial role in regulating our planet's climate and supporting life. Understanding the composition and structure of the atmosphere is essential for understanding the impacts of human activities on the planet and developing strategies for mitigating these impacts.

Chapter 4: The Earth's Oceans

The Earth's oceans cover approximately 71% of the planet's surface and play a crucial role in regulating the Earth's climate and supporting life. In this chapter, we will explore the structure and properties of the Earth's oceans, as well as their importance to the planet and to human society.

Composition and Structure of the Oceans

The Earth's oceans are composed of saltwater, with an average salinity of approximately 35 parts per thousand. The oceans are divided into five major basins: the Atlantic, Pacific, Indian, Southern, and Arctic oceans.

The structure of the ocean is divided into several layers based on temperature and salinity changes with depth. The surface layer is the warmest and the most variable, and it is where most marine life and weather occur. Below the surface layer is the thermocline, where the temperature drops rapidly with depth. Below the thermocline is the deep ocean, where the temperature is cold and the pressure is high.

Properties of the Oceans

The Earth's oceans have several unique properties that make them essential to life on the planet. Some of the key properties of the oceans include:

1. **Absorption of Solar Energy:** The oceans absorb a significant amount of solar energy, helping to regulate the Earth's temperature and climate.
2. **Carbon Dioxide Sink:** The oceans absorb a significant amount of carbon dioxide from the atmosphere, helping to mitigate the impacts of

human activities on the planet's climate.

3. **Source of Food and Resources:** The oceans support a diverse range of marine life, providing food and resources for human societies around the world.

4. **Transportation and Trade:** The oceans are a major transportation route, supporting international trade and commerce.

5. **Recreation and Tourism:** The oceans also provide opportunities for recreation and tourism, with activities such as swimming, surfing, and scuba diving.

Human Impacts on the Oceans

Human activities have had a significant impact on the health and well-being of the Earth's oceans. Some of the key impacts of human activities on the oceans include:

1. **Overfishing:** Overfishing has depleted fish populations around the world, leading to the collapse of some fisheries.

2. **Pollution:** Pollution from human activities such as plastic waste, oil spills, and nutrient runoff has harmed marine life and degraded water quality in some areas.

3. **Climate Change:** Climate change is causing ocean temperatures to rise, leading to coral bleaching and changes in marine ecosystems.

4. **Acidification:** The absorption of carbon dioxide by the oceans is causing the pH of seawater to decrease, leading to acidification that can harm marine life.

5. **Coastal Development:** Development along

coastlines can lead to the destruction of important coastal habitats and alter water flows.

Conclusion

The Earth's oceans are a vital component of our planet, playing a crucial role in regulating the Earth's climate and supporting life. However, human activities have had a significant impact on the health and well-being of the oceans, threatening their ability to support life and human societies. Understanding the importance of the oceans and the impacts of human activities is essential for developing strategies to protect and sustainably manage this critical resource.

Chapter 5: The Earth's Geology

Geology is the study of the Earth's physical structure, composition, and history. The study of geology helps us to understand how the Earth was formed, how it has changed over time, and how its natural resources can be utilized. In this chapter, we will explore the various components of the Earth's geology, including its structure, tectonic plates, rocks, minerals, and natural resources.

Structure of the Earth

The Earth's interior is divided into several layers, including the crust, mantle, outer core, and inner core. The crust is the outermost layer and is composed of solid rock, with an average thickness of about 30 kilometers on land and 5 kilometers beneath the oceans. The mantle is the largest layer and is composed of hot, molten rock. The outer core is liquid, while the inner core is solid and composed mostly of iron.

Tectonic Plates

The Earth's crust is not one solid piece, but is instead broken into a series of large plates that move and interact with each other. These tectonic plates are responsible for many of the Earth's geological features, including mountains, earthquakes, and volcanoes. The movement of the plates is driven by convection currents in the mantle, as hotter material rises and cooler material sinks.

Rocks and Minerals

Rocks are composed of minerals, which are naturally occurring substances with a defined chemical composition and crystal structure. There are three types of rocks: igneous, sedimentary, and metamorphic. Igneous rocks form from the cooling and solidification of magma or lava, while sedimentary rocks are formed from the accumulation of sediment over time. Metamorphic rocks form from the alteration of existing rocks due to heat and pressure.

Natural Resources

The Earth's geology provides us with a wide range of natural resources, including minerals, fossil fuels, and groundwater. Minerals are used in a variety of industries, including construction, electronics, and transportation. Fossil fuels, including coal, oil, and natural gas, provide energy for electricity and transportation. Groundwater is a critical resource for drinking water and agriculture.

Impacts of Human Activities on Geology

Human activities have had a significant impact on the Earth's geology. Mining, drilling, and other resource extraction activities can cause significant environmental damage, including soil erosion, water pollution, and habitat destruction. Climate change is also affecting the Earth's geology, with rising temperatures causing permafrost to thaw, leading to landslides and changes in the stability of mountains.

Conclusion

Geology is a complex and fascinating field that helps us to understand the Earth's physical structure, composition, and history. By studying the Earth's geology, we can better understand the natural processes that shape our planet, as well as the impacts of human activities on the Earth's natural resources and environment. As we continue to rely on the Earth's natural resources for energy, construction, and other industries, it is essential that we take a responsible approach to

16

resource extraction and management to protect the health and sustainability of our planet.

Chapter 6: The Earth's Climate

Climate is the long-term average of weather patterns in a particular region. The Earth's climate is influenced by a variety of factors, including the Earth's orbit and tilt, the amount of solar radiation received, the Earth's atmospheric composition, and natural and human-induced changes. In this chapter, we will explore the different factors that influence the Earth's climate, the science behind climate change, and the impacts of climate change on our planet.

Factors Influencing the Earth's Climate

The Earth's climate is influenced by a variety of natural factors, including solar radiation, volcanic activity, and changes in the Earth's orbit and tilt. Solar radiation drives the Earth's climate, as it provides the energy that drives the Earth's atmospheric and oceanic circulations. Volcanic activity can also influence the Earth's climate, as it releases large amounts of greenhouse gases into the atmosphere. Changes in the Earth's orbit and tilt can cause fluctuations in the amount of solar radiation received by the Earth, leading to changes in the Earth's climate over long periods of time.

Greenhouse Effect and Climate Change

The Earth's atmosphere is composed of a mixture of gasses, including nitrogen, oxygen, and trace amounts of greenhouse gasses such as carbon dioxide, methane, and water vapor. These gasses trap heat in the Earth's atmosphere, which helps to regulate the Earth's temperature and keep it within a range that is suitable for life. However, human activities such as the burning of fossil fuels, deforestation, and agriculture have increased the concentration of greenhouse gasses in the atmosphere, leading to an enhanced greenhouse effect and causing the Earth's temperature to rise.

Impacts of Climate Change

Climate change has a wide range of impacts on the Earth's environment and ecosystems. Rising temperatures can lead to more frequent and intense heat waves, droughts, and wildfires, while changes in precipitation patterns can lead to flooding and water scarcity. Sea levels are also rising due to the melting of glaciers and ice caps, which can lead to the loss of coastal habitats and increased flooding in coastal communities. Changes in the Earth's climate can also impact agriculture, wildlife, and human health.

Mitigating Climate Change

Reducing greenhouse gas emissions is essential to mitigating the impacts of climate change. This can be achieved through a variety of measures, including the use of renewable energy sources, improving energy efficiency, and reducing deforestation and land use changes. International agreements such as the Paris Agreement aim to limit the rise in global temperature to well below 2 degrees Celsius above pre-industrial levels, and to pursue efforts to limit the temperature increase to 1.5 degrees Celsius.

Conclusion

The Earth's climate is a complex system that is influenced by a variety of natural and human-induced factors. Climate change is one of the most pressing challenges facing our planet today, with significant impacts on our environment, ecosystems, and human societies. It is essential that we take action to reduce greenhouse gas emissions and mitigate the impacts of climate change in order to protect the health and sustainability of our planet for future generations.

Chapter 7: The Earth's Ecosystems

An ecosystem is a community of living organisms and their nonliving environment, including the physical and chemical factors that influence the organisms and their interactions with each other. The Earth is home to a vast array of ecosystems, ranging from tropical rainforests to arctic tundras. In this chapter, we will explore the different types of ecosystems on Earth, the biodiversity they support, and the important role they play in the Earth's ecology.

Types of Ecosystems

Ecosystems can be classified into different types based on their climate, vegetation, and geography. Some of the major types of ecosystems include forests, grasslands, deserts, tundras, wetlands, and aquatic ecosystems such as oceans, lakes, and rivers. Each of these ecosystems is unique in terms of the organisms that inhabit them, the ecological processes that occur within them, and the services they provide to humans.

Biodiversity

Biodiversity refers to the variety of life on Earth, including the diversity of species, genes, and ecosystems. Ecosystems with high levels of biodiversity are more resilient to environmental changes, as they are better able to adapt to changing conditions. The Earth's ecosystems support a vast array of species, ranging from microscopic bacteria to large mammals such as elephants and whales. However, human activities such as habitat destruction, pollution, and climate change are leading to a decline in biodiversity, which can have serious ecological and economic consequences.

Ecological Processes

Ecological processes are the interactions between organisms and their environment, including the cycling of nutrients, the flow of energy, and the maintenance of ecological relationships such as predator-prey interactions and symbiotic relationships. These processes are essential for the functioning of ecosystems and the services they provide, such as nutrient cycling, water purification, and carbon sequestration.

Ecosystem Services

Ecosystem services are the benefits that humans derive from ecosystems, including provisioning services such as food, water, and timber, regulating services such as climate regulation, disease control, and water purification, and cultural services such as recreational and aesthetic values. Ecosystem services are essential for human well-being and economic development, but they are often undervalued and threatened by human activities such as land use changes and pollution.

Conservation and Restoration

Conservation and restoration are important strategies for protecting the Earth's ecosystems and biodiversity. Conservation involves the preservation and management of ecosystems and their biodiversity, while restoration involves the rehabilitation of degraded ecosystems to their original state. Conservation and restoration can be achieved through a variety of measures, including protected areas, sustainable land use practices, and the restoration of degraded ecosystems.

Conclusion

The Earth's ecosystems are essential for the survival and well-being of all living organisms, including humans. They support a vast array of biodiversity, ecological processes, and ecosystem services, but they are threatened by human activities such as habitat destruction, pollution, and climate change. It is essential that we take action to protect and restore the Earth's ecosystems in order to ensure the sustainability of our planet for future generations.

Chapter 8: The Water Cycle

The water cycle, also known as the hydrological cycle, describes the continuous movement of water on, above, and below the surface of the Earth. The water cycle is a crucial component of the Earth's environmental system, as it plays a vital role in the distribution of water across the planet, the regulation of the Earth's climate, and the provision of freshwater resources for human use. In this chapter, we will explore the different processes involved in the water cycle and the importance of the water cycle for the Earth's ecosystems and human society.

The Processes of the Water Cycle

The water cycle consists of a series of processes that work together to move water around the Earth. These processes include evaporation, transpiration, condensation, precipitation, infiltration, runoff, and groundwater flow. Evaporation is the process by which water changes from a liquid to a gas, primarily from the surface of oceans, lakes, and rivers. Transpiration is the process by which water is released into the atmosphere by plants. Condensation is the process by which water vapor in the atmosphere is converted into liquid droplets. Precipitation is the process by which water falls from the atmosphere as rain, snow, or hail. Infiltration is the process by which water enters the soil and moves into underground aquifers. Runoff is the movement of water over the surface of the Earth and into rivers, lakes, and oceans. Groundwater flow is the movement of water through underground aquifers.

The Importance of the Water Cycle

The water cycle is essential for the Earth's ecosystems and human society. It plays a vital role in regulating the Earth's climate, as water vapor is a potent greenhouse gas that helps to trap heat in the atmosphere. The water cycle also helps to distribute freshwater resources across the planet, providing vital resources for agriculture, industry, and human consumption. In addition, the water cycle is essential for the functioning of the Earth's ecosystems, as it supports the growth of plants, the survival of aquatic organisms, and the maintenance of wetlands and other important habitats.

Impacts of Human Activities on the Water Cycle

Human activities have a significant impact on the water cycle, with consequences for the Earth's ecosystems and human society. Climate change, for example, is altering the Earth's hydrological cycle, leading to changes in precipitation patterns, melting of glaciers and ice caps, and changes in the timing and intensity of floods and droughts. Human activities such as deforestation, urbanization, and agriculture also affect the water cycle by altering the amount and distribution of vegetation, increasing the amount of impervious surfaces, and increasing the use of water resources for irrigation and other purposes.

Managing the Water Cycle

Managing the water cycle is essential for ensuring the sustainability of freshwater resources and protecting the Earth's ecosystems. This can be achieved through a range of measures, including water conservation and efficiency, land use management, and the development of sustainable water infrastructure such as rainwater harvesting, wastewater treatment, and groundwater management. In addition, international cooperation is needed to address transboundary water issues, such as the management of shared rivers and aquifers.

Conclusion

The water cycle is a complex and essential component of the Earth's environmental system. It plays a vital role in the distribution of water across the planet, the regulation of the Earth's climate, and the provision of freshwater resources for human use. However, the water cycle is under threat from human activities such as climate change, deforestation, and overuse of water resources. It is essential that we take action to manage the water cycle in a sustainable and equitable manner to ensure the long-term health of the Earth's ecosystems and human society.

Chapter 9: The Carbon Cycle

The carbon cycle is a fundamental component of the Earth's environmental system, describing the movement of carbon through various forms in the biosphere, atmosphere, hydrosphere, and geosphere. The carbon cycle plays a vital role in regulating the Earth's climate and supporting the growth of plants, which are the foundation of the Earth's food webs. In this chapter, we will explore the different processes involved in the carbon cycle, the role of humans in altering the carbon cycle, and the impacts of these changes on the Earth's ecosystems and human society.

The Processes of the Carbon

Cycle The carbon cycle is driven by a series of processes that move carbon from one reservoir to another. These processes include photosynthesis, respiration, decomposition, fossil fuel combustion, and ocean uptake. Photosynthesis is the process by which plants use sunlight to convert carbon dioxide and water into organic compounds, such as sugars and starches. Respiration is the process by which living organisms break down organic compounds to release energy, producing carbon dioxide as a byproduct. Decomposition is the process by which dead organic matter is broken down by bacteria and fungi, releasing carbon dioxide into the atmosphere. Fossil fuel combustion is the process by which carbon stored in fossil fuels, such as coal, oil, and gas, is released into the atmosphere when these fuels are burned. Finally, ocean uptake is the process by which the ocean absorbs carbon dioxide from the atmosphere, increasing the acidity of seawater.

The Role of Humans in Altering the Carbon Cycle

Human activities have significantly altered the carbon cycle, primarily through the burning of fossil fuels and deforestation. The burning of fossil fuels releases carbon dioxide into the atmosphere, increasing the concentration of this greenhouse gas and contributing to climate change. Deforestation removes trees that absorb carbon dioxide through photosynthesis, reducing the Earth's carbon sink capacity. In addition, land use changes such as agriculture and urbanization can alter the carbon cycle by changing the amount and distribution of vegetation.

Impacts of Changes in the Carbon Cycle

Changes in the carbon cycle have significant impacts on the Earth's ecosystems and human society. Climate change, resulting from increased greenhouse gas concentrations in the atmosphere, is altering the Earth's temperature, precipitation patterns, and sea levels, leading to a range of environmental and social impacts, including the loss of biodiversity, the spread of diseases, and the displacement of human populations. Changes in the carbon cycle can also have impacts on agricultural productivity and the availability of freshwater resources, as well as the functioning of the Earth's ocean ecosystems.

Managing the Carbon Cycle

Managing the carbon cycle is essential for mitigating the impacts of climate change and ensuring the sustainability of the Earth's ecosystems and human society. This can be achieved through a range of measures, including reducing greenhouse gas emissions from fossil fuels, promoting renewable energy sources, conserving forests and other carbon sinks, and promoting sustainable land use practices. In addition, international cooperation is needed to address the global nature of the carbon cycle and to develop effective strategies for reducing greenhouse gas emissions and promoting carbon sequestration.

Conclusion

The carbon cycle is a critical component of the Earth's environmental system, regulating the Earth's climate and supporting the growth of plants. However, human activities are altering the carbon cycle, contributing to climate change and the degradation of the Earth's ecosystems. It is essential that we take action to manage the carbon cycle in a sustainable and equitable manner, to ensure the long-term health of the Earth's ecosystems and human society. By reducing greenhouse gas emissions, promoting renewable energy sources, conserving forests, and promoting sustainable land use practices, we can work towards a more sustainable and resilient future for our planet.

Chapter 10: The Nitrogen Cycle

Nitrogen is an essential element for life on Earth, playing a critical role in the structure and function of amino acids, proteins, and nucleic acids. The nitrogen cycle describes the movement of nitrogen through various forms in the biosphere, atmosphere, hydrosphere, and geosphere, and is a critical component of the Earth's environmental system. In this chapter, we will explore the different processes involved in the nitrogen cycle, the role of humans in altering the nitrogen cycle, and the impacts of these changes on the Earth's ecosystems and human society.

The Processes of the Nitrogen Cycle

The nitrogen cycle is driven by a series of processes that move nitrogen from one reservoir to another. These processes include nitrogen fixation, nitrification, denitrification, and ammonification. Nitrogen fixation is the process by which nitrogen gas in the atmosphere is converted into organic compounds, such as ammonia and nitrate, through the action of nitrogen-fixing bacteria. Nitrification is the process by which ammonia and ammonium are converted into nitrate by nitrifying bacteria. Denitrification is the process by which nitrate is converted back into nitrogen gas by denitrifying bacteria, completing the cycle. Finally, ammonification is the process by which organic nitrogen compounds, such as amino acids and nucleic acids, are broken down into ammonia by decomposing bacteria and fungi.

The Role of Humans in Altering the Nitrogen Cycle

Human activities have significantly altered the nitrogen cycle, primarily through the use of synthetic fertilizers and the burning of fossil fuels. Synthetic fertilizers are a significant source of nitrogen, providing nutrients to crops that support human populations. However, excess fertilizer use can lead to eutrophication, where nitrogen runoff leads to algal blooms and the depletion of oxygen in aquatic ecosystems. Burning fossil fuels also releases nitrogen oxides into the atmosphere, contributing to acid rain and other environmental impacts.

Impacts of Changes in the Nitrogen Cycle

Changes in the nitrogen cycle have significant impacts on the Earth's ecosystems and human society. Eutrophication, resulting from excess nitrogen runoff, can lead to the loss of biodiversity, the spread of harmful algal blooms, and the degradation of aquatic ecosystems. Nitrogen deposition from fossil fuel burning can also contribute to acid rain, which can damage forests, soil, and aquatic ecosystems. Additionally, changes in the nitrogen cycle can affect agricultural productivity, soil fertility, and greenhouse gas emissions.

Managing the Nitrogen Cycle

Managing the nitrogen cycle is essential for ensuring the sustainability of the Earth's ecosystems and human society. This can be achieved through a range of measures, including reducing excess fertilizer use, promoting sustainable agricultural practices, conserving wetlands and other natural nitrogen sinks, and reducing fossil fuel emissions. In addition, international cooperation is needed to address the global nature of the nitrogen cycle and to develop effective strategies for reducing nitrogen pollution and promoting nitrogen sequestration.

Conclusion

The nitrogen cycle is a critical component of the Earth's environmental system, playing a vital role in supporting life on Earth. However, human activities are altering the nitrogen cycle, contributing to eutrophication, acid rain, and other environmental impacts. It is essential that we take action to manage the nitrogen cycle in a sustainable and equitable manner, to ensure the long-term health of the Earth's ecosystems and human society. By reducing excess fertilizer use, promoting sustainable agricultural practices, and conserving natural nitrogen sinks, we can work towards a more sustainable and resilient future for our planet.

Chapter 11: The Phosphorus Cycle

Phosphorus is an essential element for life on Earth, playing a critical role in the structure and function of nucleic acids, cell membranes, and energy transfer molecules. The phosphorus cycle describes the movement of phosphorus through various forms in the biosphere, geosphere, and hydrosphere, and is a critical component of the Earth's environmental system. In this chapter, we will explore the different processes involved in the phosphorus cycle, the role of humans in altering the phosphorus cycle, and the impacts of these changes on the Earth's ecosystems and human society.

The Processes of the Phosphorus Cycle

The phosphorus cycle is driven by a series of processes that move phosphorus from one reservoir to another. These processes include weathering, erosion, and sedimentation, which release phosphorus from rocks and minerals, and transport it to the oceans and other aquatic ecosystems. In aquatic ecosystems, phosphorus can be taken up by aquatic plants and phytoplankton, and incorporated into the food web. When organisms die and decompose, the phosphorus is returned to the soil or water, completing the cycle.

The Role of Humans in Altering the Phosphorus Cycle

Human activities have significantly altered the phosphorus cycle, primarily through the use of synthetic fertilizers and the runoff of animal manure and sewage into aquatic ecosystems. Synthetic fertilizers are a significant source of phosphorus, providing nutrients to crops that support human populations. However, excess fertilizer use can lead to eutrophication, where phosphorus runoff leads to algal blooms and the depletion of oxygen in aquatic ecosystems. Runoff of animal manure and sewage also contributes to eutrophication.

Impacts of Changes in the Phosphorus Cycle

Changes in the phosphorus cycle have significant impacts on the Earth's ecosystems and human society. Eutrophication, resulting from excess phosphorus runoff, can lead to the loss of biodiversity, the spread of harmful algal blooms, and the degradation of aquatic ecosystems. Additionally, changes in the phosphorus cycle can affect agricultural productivity, soil fertility, and greenhouse gas emissions.

Managing the Phosphorus Cycle

Managing the phosphorus cycle is essential for ensuring the sustainability of the Earth's ecosystems and human society. This can be achieved through a range of measures, including reducing excess fertilizer use, promoting sustainable agricultural practices, conserving wetlands and other natural phosphorus sinks, and treating animal manure and sewage before it is released into aquatic ecosystems. In addition, international cooperation is needed to address the global nature of the phosphorus cycle and to develop effective strategies for reducing phosphorus pollution and promoting phosphorus sequestration.

Conclusion

The phosphorus cycle is a critical component of the Earth's environmental system, playing a vital role in supporting life on Earth. However, human activities are altering the phosphorus cycle, contributing to eutrophication and other environmental impacts. It is essential that we take action to manage the phosphorus cycle in a sustainable and equitable manner, to ensure the long-term health of the Earth's ecosystems and human society. By reducing excess fertilizer use, promoting sustainable agricultural practices, and conserving natural phosphorus sinks, we can work towards a more sustainable and resilient future for our planet.

Chapter 12: Natural Disasters

Natural disasters are events that occur naturally and can cause significant damage to property, infrastructure, and human life. These events can be caused by a variety of natural phenomena, including earthquakes, volcanic eruptions, hurricanes, floods, tsunamis, wildfires, and landslides. In this chapter, we will explore the different types of natural disasters, their causes, impacts, and methods of management and mitigation.

Types of Natural Disasters

Earthquakes are sudden and violent shaking of the ground, caused by the release of energy stored in the Earth's crust. Volcanic eruptions occur when molten rock, ash, and gas are released from a volcanic vent or fissure. Hurricanes are tropical cyclones characterized by strong winds, heavy rain, and storm surges. Floods occur when water levels rise above the normal level in a river, lake, or ocean, or due to heavy rainfall. Tsunamis are caused by underwater earthquakes or landslides and result in massive ocean waves. Wildfires are uncontrolled fires that burn natural areas, often fueled by dry vegetation and strong winds. Landslides occur when large amounts of earth and rock slide down a slope.

Causes of Natural Disasters

Natural disasters are caused by various natural phenomena, including geological processes such as tectonic plate movement, volcanic activity, and landslides, as well as atmospheric processes like hurricanes, tornadoes, and floods. Climate change is also increasing the frequency and severity of some natural disasters, such as heatwaves, wildfires, and hurricanes.

Impacts of Natural Disasters

The impacts of natural disasters can be devastating, resulting in loss of life, displacement of people, damage to infrastructure, and long-term economic and environmental impacts. The severity of the impacts depends on the intensity of the disaster, the vulnerability of the affected area, and the level of preparedness and response by governments and emergency services.

Management and Mitigation of Natural Disasters

Effective management and mitigation of natural disasters require a range of measures, including prevention, preparedness, response, and recovery. Prevention measures include land-use planning, building codes, and early warning systems. Preparedness measures involve developing emergency plans, stockpiling supplies, and training emergency responders. Response measures include search and rescue, medical care, and distribution of food, water, and shelter. Recovery measures involve rebuilding damaged infrastructure, restoring ecosystems, and providing financial assistance to affected communities.

Conclusion

Natural disasters are a part of the natural world and can have devastating impacts on human society and the environment. However, with effective management and mitigation strategies, the impacts of these disasters can be reduced. Governments, emergency services, and individuals must work together to develop and implement effective measures to prevent, prepare for, respond to, and recover from natural disasters. By doing so, we can help ensure the safety and resilience of our communities and the natural world.

Chapter 13: Climate Change and Global Warming

Climate change is one of the most significant environmental issues facing the world today. It refers to long-term changes in the Earth's climate, including changes in temperature, precipitation, and weather patterns. One of the main drivers of climate change is global warming, which is caused by the increasing concentrations of greenhouse gasses in the atmosphere, primarily carbon dioxide.

Causes of Global Warming

The primary cause of global warming is human activities, including the burning of fossil fuels (coal, oil, and gas) for energy, deforestation, and agriculture. These activities release large amounts of greenhouse gasses, primarily carbon dioxide, into the atmosphere, trapping heat and causing the Earth's temperature to rise. Other human activities, such as industrial processes, transportation, and waste disposal, also contribute to greenhouse gas emissions.

Effects of Global Warming

Global warming has significant impacts on the Earth's climate and ecosystems. Rising temperatures lead to melting glaciers and sea ice, resulting in rising sea levels. This can cause flooding and erosion, especially in low-lying coastal areas. Changes in temperature and precipitation patterns can also affect agriculture, water resources, and natural ecosystems, leading to loss of biodiversity and food insecurity. Extreme weather events such as heatwaves, droughts, and wildfires are becoming more frequent and severe due to global warming.

Mitigation and Adaptation Strategies

Mitigating global warming involves reducing greenhouse gas emissions by transitioning to renewable energy sources, improving energy efficiency, and implementing policies and regulations to limit emissions. Adaptation strategies involve planning and implementing measures to minimize the impacts of climate change on communities and ecosystems. These include building sea walls and other coastal protections, developing drought-resistant crops, and preserving and restoring natural ecosystems to help buffer against extreme weather events.

Global Cooperation and Action

Addressing global warming and climate change requires international cooperation and action. The United Nations Framework Convention on Climate Change (UNFCCC) was established in 1992 to address climate change and facilitate international cooperation on mitigation and adaptation. The Paris Agreement, adopted in 2015, aims to limit global warming to well below 2°C above pre-industrial levels and to pursue efforts to limit the temperature increase to 1.5°C. Countries have submitted national plans (known as Nationally Determined Contributions) to reduce their greenhouse gas emissions, and periodic global stocktakes assess progress towards meeting these targets.

Conclusion

Global warming and climate change are significant environmental issues that require urgent action to mitigate and adapt to their impacts. Reducing greenhouse gas emissions and transitioning to renewable energy sources are critical steps towards limiting global warming and its effects on the planet.

40

International cooperation and action are necessary to address this global challenge and ensure a sustainable future for generations to come.

Chapter 14: The Greenhouse Effect

The greenhouse effect is a natural process that occurs when certain gasses in the Earth's atmosphere, known as greenhouse gasses, trap heat from the sun and prevent it from escaping back into space. This process is essential for life on Earth, as it helps regulate the planet's temperature and keeps it within a habitable range. However, human activities, such as burning fossil fuels and deforestation, have increased the concentrations of greenhouse gasses in the atmosphere, causing the enhanced greenhouse effect and leading to global warming and climate change.

Greenhouse Gasses

The main greenhouse gasses in the Earth's atmosphere are carbon dioxide (CO_2), methane (CH_4), nitrous oxide (N_2O), and fluorinated gasses. These gasses absorb and re-emit infrared radiation, trapping heat in the Earth's atmosphere and warming the planet's surface. Carbon dioxide is the most significant greenhouse gas, accounting for about three-quarters of human-caused emissions.

Sources of Greenhouse Gas Emissions

Human activities are the primary sources of greenhouse gas emissions, primarily from burning fossil fuels (coal, oil, and gas) for energy, deforestation, and agriculture. These activities release large amounts of greenhouse gasses into the atmosphere, increasing their concentrations and leading to global warming and climate change.

Impacts of the Greenhouse Effect

The enhanced greenhouse effect caused by human activities has significant impacts on the Earth's climate and ecosystems. Rising temperatures lead to melting glaciers and sea ice, resulting in rising sea levels. This can cause flooding and erosion, especially in low-lying coastal areas. Changes in temperature and precipitation patterns can also affect agriculture, water resources, and natural ecosystems, leading to loss of biodiversity and food insecurity. Extreme weather events such as heatwaves, droughts, and wildfires are becoming more frequent and severe due to the greenhouse effect.

Mitigation Strategies

Mitigating the greenhouse effect requires reducing greenhouse gas emissions by transitioning to renewable energy sources, improving energy efficiency, and implementing policies and regulations to limit emissions. Carbon capture and storage technology can also help reduce emissions from industrial processes. Planting trees and preserving and restoring natural ecosystems can also help absorb and store carbon from the atmosphere.

Conclusion

The greenhouse effect is a natural process that is essential for life on Earth. However, human activities have increased the concentrations of greenhouse gases in the atmosphere, leading to the enhanced greenhouse effect, global warming, and climate change. Addressing this global challenge requires urgent action to reduce greenhouse gas emissions and transition to a sustainable, low-carbon future. Mitigation strategies such as transitioning to renewable energy sources, improving energy efficiency, and implementing policies and regulations can help limit greenhouse gas emissions and reduce their impacts on the planet.

Chapter 15: The Ozone Layer

The ozone layer is a region of the Earth's stratosphere that contains a high concentration of ozone (O3) molecules. This layer plays a crucial role in protecting life on Earth by absorbing harmful ultraviolet (UV) radiation from the sun. However, human activities have led to the depletion of the ozone layer, leading to increased UV radiation reaching the Earth's surface and posing a significant threat to human health and the environment.

Ozone Formation and Depletion

Ozone is formed naturally in the Earth's atmosphere when UV radiation interacts with oxygen (O2) molecules, splitting them into individual oxygen atoms. These atoms then combine with other O2 molecules, forming ozone (O3) molecules. However, human activities, primarily the release of chlorofluorocarbons (CFCs) and other ozone-depleting substances, have led to the depletion of the ozone layer. CFCs break down in the stratosphere, releasing chlorine and other chemicals that destroy ozone molecules, leading to the thinning of the ozone layer.

Impacts of Ozone Depletion

The depletion of the ozone layer has significant impacts on human health and the environment. Increased UV radiation can cause skin cancer, cataracts, and other health problems in humans and animals. It can also damage crops, forests, and aquatic ecosystems, leading to reduced productivity and biodiversity loss. Ozone depletion can also affect the climate, as the same gasses that deplete the ozone layer can also contribute to global warming.

Ozone Layer Protection

International efforts to protect the ozone layer have been successful in reducing the use of ozone-depleting substances. The Montreal Protocol, signed in 1987, is an international treaty designed to phase out the production and consumption of ozone-depleting substances. As a result of this treaty, the production and consumption of these substances have been significantly reduced, and the ozone layer is expected to recover by mid-century.

Conclusion

The ozone layer is a vital part of the Earth's atmosphere, playing a crucial role in protecting life on Earth by absorbing harmful UV radiation. However, human activities have led to its depletion, causing significant impacts on human health and the environment. International efforts to protect the ozone layer have been successful, and the ozone layer is expected to recover by mid-century. It is crucial to continue efforts to protect the ozone layer and prevent further damage to the Earth's atmosphere.

Chapter 16: The Effects of Air Pollution

Air pollution is a significant environmental issue that affects the health and well-being of both humans and the natural world. Air pollution can come from a variety of sources, including industrial and vehicular emissions, wildfires, and natural events such as volcanic eruptions. In this chapter, we will discuss the effects of air pollution on human health, the environment, and the economy.

Human Health Effects

Exposure to air pollution can have significant impacts on human health. The pollutants in the air can cause respiratory problems such as asthma, bronchitis, and lung cancer. Long-term exposure to air pollution has been linked to heart disease, stroke, and other cardiovascular problems. Children, the elderly, and people with preexisting health conditions are particularly vulnerable to the effects of air pollution.

Environmental Effects

Air pollution can also have significant impacts on the natural world. Pollutants such as nitrogen oxides and sulfur dioxide can cause acid rain, which damages soil, waterways, and wildlife. Ozone pollution can damage crops and forests, reducing their productivity and biodiversity. Air pollution also contributes to climate change, as greenhouse gasses trap heat in the atmosphere, leading to rising global temperatures and changes in weather patterns.

Economic Effects

The impacts of air pollution also extend to the economy. Air pollution can lead to increased healthcare costs as people suffer from respiratory and cardiovascular problems. It can also damage crops and forests, leading to reduced agricultural productivity. The effects of climate change, which is exacerbated by air pollution, can have significant economic impacts, including increased damage from extreme weather events, rising sea levels, and disruptions to global supply chains.

Preventing Air Pollution

Preventing air pollution requires a combination of individual and collective action. Individuals can take steps to reduce their own contribution to air pollution, such as reducing their energy consumption and using public transportation instead of driving. Governments can implement policies to reduce emissions from industrial and transportation sources, such as regulating emissions and incentivizing the use of clean energy. International agreements, such as the Paris Agreement, can also play a crucial role in reducing global emissions and mitigating the impacts of climate change.

Conclusion

Air pollution is a significant environmental issue that affects human health, the natural world, and the economy. The pollutants in the air can cause respiratory problems, damage the environment, and contribute to climate change. Preventing air pollution requires collective action from individuals, governments, and international organizations. It is crucial to take steps to reduce emissions and mitigate the impacts of air pollution to protect the health and well-being of both people and the planet.

Chapter 17: The Effects of Water Pollution

Water pollution is a major environmental issue that affects aquatic ecosystems, wildlife, and human health. It occurs when harmful substances are released into bodies of water, including rivers, lakes, oceans, and groundwater. In this chapter, we will discuss the effects of water pollution on aquatic life, human health, and the economy.

Aquatic Life Effects

Water pollution can have devastating effects on aquatic life. Pollutants such as heavy metals, pesticides, and industrial chemicals can accumulate in the tissues of fish and other aquatic animals, leading to a range of health problems and reproductive issues. High levels of nutrients, such as nitrogen and phosphorus, can lead to eutrophication, which can cause harmful algal blooms and deplete oxygen levels in the water, leading to the death of fish and other aquatic organisms.

Human Health Effects

Water pollution can also have significant impacts on human health. Drinking contaminated water can lead to a range of health problems, including gastrointestinal illnesses, neurological disorders, and certain types of cancer. Exposure to water pollution can also cause skin rashes and respiratory problems. Children, the elderly, and people with weakened immune systems are particularly vulnerable to the effects of water pollution.

Economic Effects

The impacts of water pollution also extend to the economy. Water pollution can lead to increased healthcare costs and lost productivity due to illness. It can also harm industries such as fisheries and tourism, which depend on clean water and healthy aquatic ecosystems. Water pollution can also impact the quality of agricultural produce, which can affect global food supply chains and lead to economic losses.

Preventing Water Pollution

Preventing water pollution requires a combination of individual and collective action. Individuals can take steps to reduce their own contribution to water pollution, such as properly disposing of hazardous waste and reducing their use of fertilizers and pesticides. Governments can implement policies to regulate industrial and agricultural practices that contribute to water pollution and incentivize the use of clean technologies. International agreements, such as the UN's Sustainable Development Goals, can also play a crucial role in reducing global water pollution.

Conclusion Water pollution is a significant environmental issue that affects aquatic ecosystems, human health, and the economy. It can harm aquatic life, cause a range of health problems, and impact industries that depend on clean water. Preventing water pollution requires collective action from individuals, governments, and international organizations. It is crucial to take steps to reduce the release of harmful substances into bodies of water and protect the health and well-being of both people and the planet.

Chapter 18: The Effects of Land Pollution

Land pollution is the degradation of land due to human activities such as industrialization, urbanization, and agricultural practices. This chapter will discuss the effects of land pollution on the environment, human health, and the economy.

Environmental Effects

Land pollution can have severe effects on the environment. It can lead to soil erosion, which reduces the ability of land to support plant life and can lead to desertification. Contaminants such as pesticides, heavy metals, and toxic chemicals can also accumulate in the soil, affecting the health of plants and animals. Land pollution can also contaminate groundwater, which is a vital source of drinking water for many people.

Human Health Effects

Land pollution can also have significant impacts on human health. Exposure to contaminated soil can lead to a range of health problems, including respiratory problems, skin rashes, and neurological disorders. Children are particularly vulnerable to the effects of land pollution due to their smaller body size and developing immune systems. The use of pesticides and other chemicals in agriculture can also pose health risks to farmworkers and nearby residents.

Economic Effects

The impacts of land pollution can also extend to the economy. Land pollution can lead to decreased crop yields, which can impact food supply chains and lead to economic losses for farmers. It can also impact industries such as real estate and tourism, as contaminated land may be less desirable for development. Additionally, cleaning up contaminated land can be a costly and time-consuming process.

Preventing Land Pollution

Preventing land pollution requires a combination of individual and collective action. Individuals can take steps to reduce their own contribution to land pollution, such as properly disposing of hazardous waste and reducing their use of pesticides and fertilizers. Governments can implement policies to regulate industrial and agricultural practices that contribute to land pollution and incentivize the use of clean technologies. International agreements, such as the Paris Agreement on climate change, can also play a crucial role in reducing global land pollution.

Conclusion

Land pollution is a significant environmental issue that affects the health of both people and the planet. It can harm the environment, lead to a range of health problems, and impact the economy. Preventing land pollution requires collective action from individuals, governments, and international organizations. It is crucial to take steps to reduce the release of harmful substances into the land and protect the health and well-being of both people and the planet.

Chapter 19: The Effects of Noise Pollution

Noise pollution is the unwanted or excessive sound that disrupts human activities and can have negative effects on health and well-being. This chapter will explore the effects of noise pollution on human health, the environment, and quality of life.

Health Effects

Exposure to noise pollution can lead to a range of health problems, including hearing loss, stress, and sleep disturbances. Prolonged exposure to high levels of noise can cause permanent hearing damage, while short-term exposure can lead to temporary hearing loss. Noise pollution can also increase stress levels, which can lead to anxiety, depression, and other mental health issues. Chronic exposure to noise pollution can disrupt sleep patterns, leading to sleep disturbances and fatigue.

Environmental Effects

Noise pollution can also have significant impacts on the environment. Excessive noise can disrupt the behavior and communication of animals, leading to habitat loss and decreased biodiversity. It can also impact the migration patterns of birds and marine mammals, causing long-term damage to their populations. Additionally, noise pollution can affect the quality of soil, water, and air, leading to ecosystem degradation.

Quality of Life

Noise pollution can also have negative effects on the quality of life for humans. It can disrupt communication, making it difficult to hear important information or hold conversations. It can also disrupt daily activities, such as studying or working. Noise pollution can also impact property values, making it difficult to sell or rent homes and reducing the desirability of living in certain areas.

Preventing Noise Pollution

Preventing noise pollution requires a combination of individual and collective action. Individuals can take steps to reduce their contribution to noise pollution by using headphones or earplugs, keeping their car mufflers in good condition, and avoiding loud activities during nighttime hours. Governments can implement regulations to limit noise emissions from sources such as transportation and industrial activities. Buildings and infrastructure can also be designed with noise reduction in mind.

Conclusion

Noise pollution is a significant problem that can have negative effects on human health, the environment, and quality of life. It is important to take steps to reduce noise pollution through individual and collective action. This includes using noise-reducing technologies, implementing regulations, and designing buildings and infrastructure with noise reduction in mind. By working together, we can reduce the impact of noise pollution and improve the health and well-being of both people and the environment.

Chapter 20: The Effects of Light Pollution

Light pollution is the excessive or misdirected artificial light that can have negative effects on human health, wildlife, and the environment. This chapter will explore the effects of light pollution and the measures that can be taken to reduce its impact.

Human Health Effects

Exposure to artificial light at night can disrupt the body's natural sleep-wake cycle, leading to health problems such as insomnia, depression, and obesity. It can also cause eye strain and headaches. Exposure to blue light, which is emitted by many electronic devices and LED lights, can also disrupt sleep and have negative effects on mental health.

Wildlife Effects

Many species of wildlife depend on natural cycles of light and darkness for survival, such as migration patterns and breeding cycles. Exposure to artificial light at night can disrupt these natural cycles and lead to confusion, disorientation, and changes in behavior. It can also attract insects and disrupt their natural patterns of activity, which can have negative effects on the food chain and ecosystem.

Environmental Effects

Light pollution can also have negative effects on the environment. It wastes energy and contributes to greenhouse gas emissions. It can also disrupt the natural balance of ecosystems by altering the behavior and distribution of species.

It can also interfere with astronomical observations and research.

Measures to Reduce Light Pollution

There are several measures that can be taken to reduce the impact of light pollution. This includes designing outdoor lighting fixtures to minimize light spill and glare, using motion sensors and timers to turn off lights when they are not needed, and choosing lighting fixtures that emit warmer colors rather than blue light. Shielding lights and directing them downwards can also help to minimize light pollution.

Conclusion

Light pollution is a growing problem that can have negative effects on human health, wildlife, and the environment. It is important to take measures to reduce light pollution through the use of proper lighting fixtures and practices. By working together, we can reduce the impact of light pollution and preserve the natural balance of ecosystems while promoting healthy sleep patterns and well-being for both humans and wildlife.

Chapter 21: Deforestation

Deforestation refers to the permanent destruction of forests or woodlands, often for the purpose of expanding agriculture, urban development, or timber production. This chapter will explore the causes, effects, and potential solutions to deforestation.

Causes of Deforestation

One of the main causes of deforestation is agriculture. As the global population continues to grow, more land is needed to grow crops and raise livestock. Logging and timber production is another major cause of deforestation, particularly in areas with valuable hardwoods such as teak and mahogany. Mining, urban development, and infrastructure projects such as roads and dams also contribute to deforestation.

Effects of Deforestation

Deforestation has significant negative effects on the environment, wildlife, and human communities. It contributes to climate change by reducing the number of trees that absorb carbon dioxide from the atmosphere. It also leads to soil erosion, loss of biodiversity, and degradation of water quality. Deforestation can also have negative impacts on human communities by disrupting traditional land-use practices, reducing access to clean water and natural resources, and increasing the risk of natural disasters such as landslides and floods.

Solutions to Deforestation

There are several strategies that can be employed to combat deforestation. One approach is to promote sustainable land-use practices, such as agroforestry and selective logging, that allow for the productive use of forest resources while minimizing environmental damage. Conservation efforts, such as the creation of protected areas and national parks, can also help to preserve critical habitats and reduce the pressure on forests. Additionally, reducing demand for products that contribute to deforestation, such as palm oil and beef, can help to reduce the economic incentives for clearing forests.

Conclusion

Deforestation is a significant environmental issue that has far-reaching impacts on the planet and its inhabitants. It is caused by a variety of factors, including agriculture, logging, mining, and infrastructure development. The effects of deforestation are numerous and include climate change, loss of biodiversity, and disruption of traditional land-use practices. However, there are several potential solutions to this problem, including sustainable land-use practices, conservation efforts, and reducing demand for products that contribute to deforestation. By taking action to address deforestation, we can help to preserve the natural world for future generations.

Chapter 22: Soil Erosion

Soil erosion is a natural process that occurs when the surface layer of soil is transported by wind or water. However, human activities such as agriculture, deforestation, and construction can accelerate erosion rates to an unsustainable level. This chapter will examine the causes, effects, and potential solutions to soil erosion.

Causes of Soil Erosion

One of the primary causes of soil erosion is land use changes, particularly those related to agriculture. When land is cleared for farming, the protective vegetation cover is removed, leaving the soil exposed to erosion by wind and water. Additionally, conventional farming practices such as tilling and monoculture can further accelerate erosion rates. Deforestation, mining, and construction activities can also contribute to soil erosion by removing vegetation and disrupting soil structure.

Effects of Soil Erosion

Soil erosion has significant negative impacts on the environment, agriculture, and human communities. It reduces soil fertility, decreases crop yields, and can lead to the loss of topsoil, which is rich in nutrients. Additionally, erosion can cause sedimentation in waterways, leading to decreased water quality and increased risk of flooding. Soil erosion can also have socio-economic impacts, particularly in developing countries where agriculture is a primary source of income. Loss of soil fertility and crop yields can exacerbate food insecurity and poverty.

Solutions to Soil Erosion

There are several strategies that can be employed to mitigate soil erosion. One approach is to promote sustainable land use practices, such as conservation tillage, cover cropping, and crop rotation, that reduce soil disturbance and maintain vegetative cover. Agroforestry, which combines trees with crops and/or livestock, can also help to reduce soil erosion rates while providing additional benefits such as improved soil fertility and carbon sequestration. Additionally, reforestation and restoration of degraded lands can help to improve soil health and reduce erosion rates.

Conclusion

Soil erosion is a significant environmental issue that has far-reaching impacts on the planet and its inhabitants. It is caused by a variety of factors, including agriculture, deforestation, and construction. The effects of soil erosion are numerous and include decreased soil fertility, decreased crop yields, and reduced water quality. However, there are several potential solutions to this problem, including sustainable land-use practices, agroforestry, and reforestation. By taking action to address soil erosion, we can help to preserve the health of our soils and ensure food security for future generations.

Chapter 23: Overfishing

Overfishing is the practice of catching fish faster than they can reproduce, leading to a decline in fish populations and negative impacts on marine ecosystems. This chapter will examine the causes, effects, and potential solutions to overfishing.

Causes of Overfishing

One of the primary causes of overfishing is the demand for seafood, driven by a growing human population and increasing wealth. As demand for seafood increases, so does the pressure on fish populations. Technological advances in fishing methods and equipment, such as sonar and trawling nets, have also made it easier to catch large quantities of fish. Additionally, subsidies and government policies that promote fishing can lead to overfishing.

Effects of Overfishing

Overfishing has significant negative impacts on marine ecosystems and human communities that depend on seafood for food and livelihoods. When fish populations decline, it can disrupt the food chain and lead to imbalances in marine ecosystems. This can have cascading effects on other species and the overall health of the ecosystem. Overfishing also has socio-economic impacts, particularly in developing countries where fishing is a primary source of income. It can lead to job losses, food insecurity, and economic instability.

Solutions to Overfishing

There are several strategies that can be employed to mitigate overfishing. One approach is to promote sustainable fishing practices, such as limiting fishing quotas and implementing fishing gear regulations to minimize bycatch. Marine protected areas can also help to protect fish populations and their habitats. Additionally, alternative protein sources such as plant-based proteins and insect protein can reduce demand for seafood. Fisheries management policies, such as catch shares and transferable fishing quotas, can also help to promote sustainable fishing practices.

Conclusion

Overfishing is a significant environmental issue that has far-reaching impacts on marine ecosystems and human communities. It is caused by a variety of factors, including demand for seafood, technological advances in fishing methods, and government policies that promote fishing. The effects of overfishing include disruptions to marine ecosystems and socio-economic impacts on fishing communities. However, there are several potential solutions to this problem, including sustainable fishing practices, marine protected areas, and alternative protein sources. By taking action to address overfishing, we can help to ensure the health of our oceans and the livelihoods of fishing communities for future generations.

Chapter 24: Conservation and Biodiversity

Biodiversity is the variety of life on Earth, encompassing the diversity of species, genes, and ecosystems. Conservation is the protection and management of biodiversity, with the goal of maintaining healthy ecosystems and sustaining natural resources. This chapter will explore the importance of biodiversity and the role of conservation in preserving it.

The Importance of Biodiversity

Biodiversity is essential for the functioning of ecosystems and the services they provide. These services include the regulation of climate, water quality, pollination, and nutrient cycling. Biodiversity also provides important resources for human use, such as food, medicine, and raw materials. Additionally, biodiversity is important for cultural and aesthetic reasons, as it enriches our lives and provides inspiration for art, literature, and music.

The Threats to Biodiversity There are several threats to biodiversity, including habitat loss and fragmentation, pollution, overexploitation of natural resources, invasive species, and climate change. Human activities, such as land-use change, urbanization, and industrialization, are major drivers of habitat loss and fragmentation. Pollution from agricultural runoff, industrial waste, and other sources can also have negative impacts on biodiversity. Overexploitation of natural resources, such as logging and overfishing, can lead to declines in biodiversity. Invasive species can outcompete native species and disrupt ecosystems. Climate change is also a significant threat to biodiversity, as it can cause shifts in species ranges, alter ecosystems, and increase the risk of extinction.

Conservation Strategies

Conservation strategies aim to protect and manage biodiversity, with the goal of maintaining healthy ecosystems and sustaining natural resources. One approach is to establish protected areas, such as national parks and wildlife reserves, which can help to preserve habitats and protect biodiversity. Habitat restoration and rewilding projects can also help to restore degraded ecosystems. Sustainable land use practices, such as agroforestry and sustainable forestry, can promote biodiversity while also supporting human livelihoods. In addition, public education and outreach can help to raise awareness about the importance of biodiversity and the need for conservation.

Conclusion

Biodiversity is essential for the functioning of ecosystems and the services they provide, as well as for human use and cultural and aesthetic reasons. However, biodiversity is threatened by a variety of factors, including habitat loss, pollution, overexploitation of natural resources, invasive species, and climate change. Conservation strategies, such as protected areas, habitat restoration, sustainable land use practices, and public education, can help to protect and manage biodiversity. By taking action to conserve biodiversity, we can help to ensure the health of our ecosystems and sustain natural resources for future generations.

Chapter 25: Endangered Species

The Earth is home to an incredibly diverse range of plant and animal species, each with their own unique characteristics and ecological niches. Unfortunately, human activities have led to a significant increase in the number of species facing extinction. In fact, some estimates suggest that up to one million species may be at risk of extinction in the coming decades. This chapter will explore the causes of species endangerment and the efforts being made to protect endangered species.

Causes of Endangerment

There are numerous factors that contribute to species endangerment. One of the most significant is habitat destruction. As human populations grow and expand, they often encroach on natural habitats, leading to the destruction of forests, grasslands, wetlands, and other ecosystems. This destruction can lead to the displacement and even extinction of species that depend on those habitats for survival.

Another major cause of species endangerment is overexploitation. This occurs when humans hunt, fish, or harvest species at a rate that exceeds their ability to reproduce. This can lead to the collapse of entire populations and, ultimately, extinction.

Climate change is also having a significant impact on species endangerment. As temperatures rise, ecosystems are being disrupted, and many species are struggling to adapt. For example, rising sea levels are leading to the loss of coastal habitats, which are home to numerous species, while changing weather patterns are affecting the timing of migration and breeding cycles.

Efforts to Protect Endangered Species

There are numerous efforts underway to protect endangered species and prevent their extinction. One of the most important is habitat preservation. This involves identifying and protecting areas of land and water that are critical to the survival of endangered species. In many cases, this involves working with governments, NGOs, and local communities to establish protected areas such as national parks, wildlife reserves, and marine sanctuaries.

Another important strategy for protecting endangered species is captive breeding. This involves breeding endangered species in captivity and releasing them back into the wild to boost their numbers. This strategy has been successfully used to save numerous species, including the California condor, the black-footed ferret, and the Arabian oryx.

In addition, numerous laws and regulations have been enacted to protect endangered species. In the United States, the Endangered Species Act is a federal law that provides for the conservation of endangered and threatened species and the ecosystems on which they depend. Similar laws and regulations exist in other countries around the world.

Finally, public education and outreach are critical to protecting endangered species. By raising awareness about the importance of biodiversity and the threats facing endangered species, individuals can take actions to reduce their impact on the environment and support efforts to protect endangered species.

Conclusion

Endangered species face numerous threats, including habitat destruction, overexploitation, and climate change. However, there are numerous efforts underway to protect endangered species and prevent their extinction. By preserving critical habitats, engaging in captive breeding programs, enacting laws and regulations, and raising public awareness, we can work to protect endangered species and ensure that they continue to thrive in the wild.

Chapter 26: Human Population Growth

Introduction

Human population growth is the increase in the number of individuals in a population over time. It is a major environmental issue as the demand for resources increases with the growing population. In this chapter, we will discuss the causes and effects of human population growth and its impact on the environment.

Causes of Human Population Growth

Human population growth can be attributed to a variety of factors such as improved healthcare, sanitation, and nutrition, as well as technological advancements in agriculture and industry. These factors have led to longer lifespans and reduced infant mortality rates, resulting in a larger overall population.

Effects of Human Population Growth on the Environment

Human population growth has a significant impact on the environment. As the population grows, so does the demand for resources such as food, water, and energy. This demand can lead to deforestation, habitat loss, and biodiversity loss. The increase in energy consumption can also lead to air pollution and greenhouse gas emissions, contributing to climate change. Additionally, human activities such as mining, agriculture, and transportation can lead to soil erosion and water pollution.

Sustainable Population Growth

Sustainable population growth is a balance between population size and resource availability. Achieving sustainable population growth requires a focus on reducing resource consumption and promoting environmentally friendly practices. This can be accomplished through education and awareness campaigns, as well as policies and regulations that promote sustainable development.

Family Planning

Family planning is an important aspect of achieving sustainable population growth. Providing access to contraception and family planning services can help to reduce population growth rates. Education and awareness campaigns can also help to promote smaller family sizes.

Urbanization

Urbanization can also play a role in reducing population growth rates. As populations become more urbanized, fertility rates tend to decrease. Additionally, urban areas are often more efficient in their use of resources, which can help to reduce the overall demand for resources.

Conclusion

Human population growth is a major environmental issue that has significant impacts on the environment. Achieving sustainable population growth requires a focus on reducing resource consumption and promoting environmentally friendly practices. Family planning and urbanization can also play important roles in achieving sustainable population growth.

Chapter 27: Renewable Energy Sources

Introduction

Renewable energy sources are sources of energy that are replenished naturally and can be used without depleting the resource. They are an important part of the transition to a sustainable energy future. In this chapter, we will discuss the different types of renewable energy sources, their advantages and disadvantages, and their potential for use in the future.

Types of Renewable Energy Sources

1. **Solar Energy** - Solar energy is the most abundant source of renewable energy on earth. It is harnessed through the use of photovoltaic cells that convert sunlight into electricity.
2. **Wind Energy** - Wind energy is harnessed through the use of wind turbines. As wind passes through the blades of the turbine, it causes them to spin, generating electricity.
3. **Hydro Energy** - Hydro energy is harnessed through the use of dams or other structures that capture the kinetic energy of moving water and convert it into electricity.
4. **Geothermal Energy** - Geothermal energy is harnessed by using the natural heat of the earth's core. It is commonly used to heat buildings and produce electricity.
5. **Biomass Energy** - Biomass energy is derived from organic matter such as wood, crops, and animal

waste. It is commonly used to produce electricity or heat.

Advantages and Disadvantages of Renewable Energy Sources

Renewable energy sources offer many advantages over non-renewable sources such as coal, oil, and gas. They are a clean and sustainable source of energy that do not emit greenhouse gasses or other pollutants. They also have the potential to create jobs and boost local economies.

However, renewable energy sources also have some disadvantages. They can be more expensive to produce and may require significant investments in infrastructure. Some renewable energy sources such as wind and solar are intermittent, meaning they are dependent on weather conditions and cannot provide a constant source of power.

Potential for Use in the Future

Renewable energy sources have the potential to play a significant role in the future of energy production. As technology continues to advance and become more efficient, the cost of producing renewable energy is expected to decrease. In addition, governments and businesses around the world are increasingly committing to reducing their carbon emissions and transitioning to renewable energy sources.

Conclusion

Renewable energy sources offer a clean and sustainable alternative to non-renewable sources of energy. While they do have some disadvantages, the potential benefits of renewable energy sources make them a critical part of the transition to a sustainable energy future. Governments and businesses should continue to invest in the development of renewable energy technology to accelerate the transition to a clean energy future.

Chapter 28: Solar Energy

Solar energy is a type of renewable energy that is becoming increasingly popular around the world. It is a clean, abundant and reliable source of energy that can be harnessed using a variety of technologies. Solar energy is derived from the sun, which is a giant nuclear reactor that constantly emits energy in the form of light and heat. This energy can be captured and converted into usable electricity or heat using a variety of methods.

History of Solar Energy

The use of solar energy dates back thousands of years, with early civilizations using the sun to heat water for bathing and cooking. The ancient Greeks and Romans built south-facing buildings and used mirrors to reflect sunlight into their homes. In the 19th century, scientists began to develop photovoltaic (PV) technology, which allowed the conversion of sunlight directly into electricity. The first solar cell was developed by Charles Fritts in 1883, and over the years, scientists and engineers have continued to refine and improve upon this technology.

Types of Solar Energy

There are two main types of solar energy: photovoltaic (PV) and concentrated solar power (CSP).

Photovoltaic (PV) systems convert sunlight directly into electricity using solar cells. These cells are made of semiconductor materials, such as silicon, and are arranged in a panel. When sunlight hits the panel, it creates an electrical current that can be used to power homes and businesses. PV systems can be installed on rooftops, on the ground, or even on top of water.

Concentrated solar power (CSP) systems use mirrors or lenses to focus sunlight onto a small area. This concentrated sunlight heats a fluid, such as water, which is then used to produce steam. The steam can be used to generate electricity using a turbine. CSP systems are typically used in large-scale power plants, and can be more efficient than PV systems in areas with high levels of direct sunlight.

Benefits of Solar Energy

There are numerous benefits to using solar energy, including:

1. **Renewable:** Solar energy is a renewable resource, which means it will never run out. The sun is expected to last for billions of years, so we can continue to rely on it as a source of energy.
2. **Clean:** Solar energy produces no emissions or pollution, making it one of the cleanest sources of energy available. This helps to reduce our carbon footprint and protect the environment.
3. **Cost-effective:** The cost of solar panels has decreased significantly over the past decade, making it more affordable for homeowners and businesses to install them. Additionally, solar energy can help to lower electricity bills and provide a source of income through net metering.
4. **Versatile:** Solar energy can be used in a variety of applications, from powering homes and businesses to providing electricity in remote areas and on

74

spacecraft.

5. **Job Creation:** The solar energy industry has created thousands of jobs worldwide, from manufacturing and installation to sales and marketing.

Challenges of Solar Energy

While solar energy has numerous benefits, there are also some challenges to its adoption and use:

1. **Intermittent:** Solar energy is an intermittent source of energy, which means it only generates electricity when the sun is shining. This can make it difficult to rely on as a sole source of energy.

2. **Storage:** Energy storage is a challenge for solar energy, as batteries and other storage technologies can be expensive and bulky. However, research and development in this area is ongoing.

3. **Land Use:** Large-scale solar energy projects can require significant amounts of land, which can impact wildlife habitats and other natural resources.

4. **Weather-dependent:** Solar energy generation can be impacted by weather conditions such as clouds and rain, which can reduce the efficiency of solar panels.

Conclusion

In conclusion, solar energy is a promising renewable energy source that has the potential to play a significant role in meeting our energy needs in the future. With the advancements in technology and decreasing costs, solar energy is becoming increasingly accessible to individuals and businesses. The benefits of solar energy, including its clean and renewable nature, cost-effectiveness, versatility, and job creation, make it an attractive option for addressing the challenges of climate change and energy security. However, there are still challenges that need to be addressed, such as the intermittency and storage of solar energy and the impact of large-scale solar projects on the environment. With ongoing research and development, solar energy has the potential to become a key component of our sustainable energy future.

Chapter 29: Wind Energy

Introduction

Energy is an essential aspect of human life. Over the years, humans have relied on fossil fuels for their energy needs. However, due to the harmful effects of fossil fuels on the environment, renewable energy sources such as wind energy have gained popularity. Wind energy is an excellent alternative to fossil fuels because it is abundant, sustainable, and emits zero greenhouse gasses. This chapter will focus on wind energy, including its history, technology, advantages, disadvantages, and future prospects.

History of Wind Energy

Wind energy has been used for centuries, primarily for grinding grain and pumping water. The earliest known use of wind energy was in Persia, where windmills were used for irrigation as early as the 7th century. Windmills were also used in Europe in the 12th century for grinding grain. With the invention of the steam engine and the discovery of fossil fuels, wind energy fell out of favor in the late 19th century.

However, in the 1970s, the oil crisis prompted renewed interest in wind energy. Since then, the technology has improved significantly, making wind energy a viable source of electricity for both onshore and offshore applications.

Technology

Wind turbines are the most common technology used to harness wind energy. They consist of blades, a rotor, a nacelle, and a tower. The blades are designed to capture the kinetic energy of the wind, which rotates the rotor. The rotor is connected to a generator in the nacelle, which converts the rotational energy into electricity. The tower supports the nacelle and blades, allowing the turbine to capture more wind energy at higher altitudes.

Advantages of Wind Energy

There are several advantages to using wind energy. Firstly, it is a renewable energy source, meaning that it is abundant and will never run out. Secondly, wind energy is sustainable and emits no greenhouse gasses, which reduces the impact of climate change. Thirdly, wind turbines can be placed on land or offshore, which makes them a flexible source of energy. Fourthly, wind energy is cost-competitive with fossil fuels in many regions, especially when considering the long-term environmental and economic costs of fossil fuels.

Disadvantages of Wind Energy

Despite its advantages, wind energy also has some disadvantages. Firstly, wind turbines can be noisy, which can be a problem for people living nearby. Secondly, wind turbines can be a hazard to birds and bats, especially during migration. Thirdly, wind turbines can be visually intrusive, which can impact the aesthetics of natural landscapes. Fourthly, wind energy is intermittent, which means that it cannot provide a constant source of energy, especially during calm weather conditions.

Future Prospects

Wind energy is a rapidly growing industry, with significant potential for expansion. As technology improves, wind turbines are becoming more efficient and cost-effective, making wind energy more competitive with fossil fuels. Offshore wind energy has significant potential for growth, as offshore wind turbines can be larger and capture more wind energy than onshore turbines. Furthermore, the integration of wind energy with other renewable energy sources, such as solar and hydro, can help address the intermittent nature of wind energy.

Conclusion

Wind energy is a promising alternative to fossil fuels, with several advantages and a few disadvantages. As the world becomes more aware of the need to transition to renewable energy sources, wind energy will play an increasingly critical role in meeting global energy needs. However, it is essential to continue to address the challenges associated with wind energy and work towards making it a more efficient and cost-effective source of energy.

Chapter 30: Hydroelectric Energy

Hydroelectric energy, also known as hydro power, is a form of renewable energy that utilizes the flow of water to generate electricity. It is a clean, reliable, and sustainable source of energy that does not produce greenhouse gas emissions or other pollutants, making it an attractive option for meeting the world's growing energy demands.

Hydroelectric power plants typically consist of a dam or reservoir, a turbine, and a generator. Water is held back by the dam, creating a reservoir that stores potential energy. When the water is released from the dam, it flows through a turbine, which converts the kinetic energy of the water into mechanical energy that rotates the turbine blades. The rotating turbine then drives a generator, which converts the mechanical energy into electrical energy.

There are several types of hydroelectric power plants, including:

1. **Conventional hydroelectric plants:** These plants use dams to create a reservoir of water that can be released to generate electricity.
2. **Pumped-storage hydroelectric plants:** These plants use excess electricity to pump water from a lower reservoir to an upper reservoir. The water can then be released to generate electricity when demand is high.
3. **Run-of-river hydroelectric plants:** These plants do not require a dam or reservoir. Instead, they use the natural flow of a river to generate electricity.

Hydroelectric power is the largest source of renewable energy worldwide, accounting for over 16% of global electricity production. It is also the most efficient form of renewable energy, with an average efficiency of around 90%.

In addition to being a clean and reliable source of energy, hydroelectric power also provides several other benefits:

1. **Flood control:** Dams can be used to regulate water flow, which can help prevent flooding during heavy rains or snowmelt.
2. **Irrigation:** Water stored in reservoirs can be used for irrigation, which can increase crop yields and support agricultural development.
3. **Recreation:** Reservoirs created by dams can be used for recreational activities such as swimming, boating, and fishing.

However, hydroelectric power also has some drawbacks. The construction of large dams can have significant environmental impacts, including altering river ecosystems, displacing communities, and disrupting natural water flow. Dams can also trap sediment, which can affect downstream habitats and fisheries. Additionally, the construction and maintenance of hydroelectric power plants can be expensive.

Despite these drawbacks, hydroelectric power remains a critical source of renewable energy and is expected to play a key role in meeting the world's growing energy demands in the coming decades.

Chapter 31: Geothermal Energy

Geothermal energy is a type of renewable energy that is generated from the heat that is naturally produced by the Earth's core. This heat is then transferred to the Earth's surface, where it can be used to generate electricity or to heat buildings. Geothermal energy is considered to be a clean and sustainable source of energy because it does not produce greenhouse gas emissions or other harmful pollutants that can harm the environment.

Geothermal energy has been used for thousands of years, dating back to the ancient Roman times when they used hot springs to heat their homes and bathe. Today, geothermal energy is used in a variety of applications, including electricity generation, heating and cooling buildings, and industrial processes.

Geothermal resources are found throughout the world, but they are most commonly found in areas that are near tectonic plate boundaries or active volcanic areas. In these areas, the Earth's crust is thin, allowing for easier access to the geothermal heat.

There are three main types of geothermal power plants: dry steam, flash steam, and binary cycle. Each type of plant uses a different method to extract and use geothermal energy.

Dry steam power plants use steam that is directly produced from the geothermal reservoir to generate electricity. This type of plant is the oldest and simplest technology and is used primarily in areas where there is a high temperature steam resource.

Flash steam power plants use hot water that is pumped from the geothermal reservoir and brought to the surface under pressure. The pressure is then released, which causes the water to flash into steam. The steam is then used to drive a turbine, which generates electricity.

Binary cycle power plants use a heat exchanger to transfer heat from the geothermal water to a secondary fluid with a lower boiling point. The secondary fluid vaporizes, which drives a turbine to generate electricity.

Geothermal energy also has applications in heating and cooling buildings. In areas with geothermal resources, heat can be extracted from the Earth and used to warm buildings in the winter. During the summer, the process can be reversed, and heat can be pumped back into the Earth to cool the building.

Despite the many benefits of geothermal energy, there are also some challenges associated with its use. The upfront costs of building geothermal power plants and drilling wells can be high, and there are also concerns about the potential for induced seismicity and the release of greenhouse gasses from geothermal reservoirs.

However, as technology continues to advance and the demand for clean energy sources increases, geothermal energy is becoming an increasingly attractive option for meeting our energy needs in a sustainable and environmentally friendly way.

Chapter 32: Biomass Energy

Biomass energy refers to the energy that is produced from organic materials, such as wood, crops, agricultural waste, and other forms of plant and animal matter. Biomass energy is renewable because it is derived from living or recently dead organisms and can be replenished over time through natural processes. Biomass energy has been used by humans for thousands of years, but it is only in recent years that it has become a significant source of renewable energy.

In this chapter, we will explore the different types of biomass energy, their advantages and disadvantages, and their potential to provide a sustainable source of energy for the future.

Types of Biomass Energy

There are several different types of biomass energy, including:

1. **Wood Energy:** Wood has been used as a source of energy for thousands of years. It is still widely used in many parts of the world for cooking and heating. Wood can be burned in stoves, fireplaces, or boilers to produce heat and electricity.

2. **Biofuels:** Biofuels are liquid or gaseous fuels that are derived from biomass. Examples of biofuels include ethanol, biodiesel, and biogas. Ethanol is produced by fermenting sugars or starches from crops such as corn or sugarcane. Biodiesel is produced from vegetable oils or animal fats. Biogas is produced by the anaerobic digestion of organic materials such as

agricultural waste, food waste, or sewage.

3. **Biopower:** Biopower refers to the production of electricity from biomass. This can be done through the burning of biomass in a boiler to produce steam, which drives a turbine to generate electricity. Biopower can also be produced through the gasification of biomass, where it is heated in the absence of oxygen to produce a synthetic gas that can be burned to generate electricity.

Advantages of Biomass Energy

1. **Renewable:** Biomass is a renewable source of energy because it can be replenished over time through natural processes such as photosynthesis.

2. **Carbon Neutral:** Biomass energy is considered to be carbon neutral because the carbon dioxide released during combustion is offset by the carbon dioxide absorbed by plants during growth.

3. **Reduced Landfill Waste:** Using biomass energy can reduce the amount of organic waste sent to landfills, which can produce methane, a potent greenhouse gas.

4. **Job Creation:** The use of biomass energy can create jobs in the agriculture and forestry sectors, as well as in the production and installation of biomass energy technologies.

Disadvantages of Biomass Energy

1. **Land Use:** The production of biomass energy requires land, which can compete with other land

uses such as food production or conservation.

2. **Emissions:** The burning of biomass can produce air pollutants such as particulate matter, nitrogen oxides, and sulfur dioxide.

3. **Cost:** Biomass energy technologies can be expensive to install and maintain, which can make them less economically viable than other forms of renewable energy.

4. **Sustainability:** The sustainability of biomass energy depends on how it is produced and harvested. Unsustainable practices such as clear-cutting forests or using large amounts of water for irrigation can have negative impacts on ecosystems and the availability of freshwater.

Conclusion

Biomass energy is a promising source of renewable energy that can help reduce dependence on fossil fuels and mitigate climate change. However, like all energy sources, it has its advantages and disadvantages, and its sustainability depends on how it is produced and used. By carefully considering the environmental and social impacts of biomass energy, we can work towards a future in which it plays a key role in the transition to a more sustainable energy system.

Chapter 33: Nuclear Energy

Nuclear energy is a type of energy that is generated through the nuclear reactions that take place in the core of an atom. It is one of the most powerful sources of energy known to humans, capable of producing a vast amount of electricity with relatively low amounts of fuel.

Nuclear energy is generated through a process called nuclear fission, where the nucleus of an atom is split into two smaller nuclei, releasing a large amount of energy in the process. This process is typically carried out in nuclear power plants, where the energy released from nuclear fission is used to generate electricity.

Advantages of Nuclear Energy

1. **Reliability:** Nuclear power plants can operate continuously for long periods of time, providing a reliable source of electricity to power homes and businesses.
2. **Low carbon emissions:** Unlike traditional fossil fuel power plants, nuclear power plants do not emit greenhouse gasses such as carbon dioxide, making them a cleaner energy source.
3. **Efficiency:** Nuclear power plants are highly efficient, able to generate large amounts of electricity using relatively small amounts of fuel.
4. **Reduced dependence on foreign oil:** Because nuclear energy is generated using uranium, a domestic source of fuel, it can help to reduce the United States' dependence on foreign oil.

Disadvantages of Nuclear Energy

1. **Radioactive waste:** Nuclear power plants generate radioactive waste, which can remain hazardous for thousands of years. Proper disposal of this waste is critical to prevent harm to the environment and human health.
2. **Safety concerns:** The possibility of nuclear accidents, such as the 1986 Chernobyl disaster, raises concerns about the safety of nuclear energy.
3. **High initial cost:** The construction of nuclear power plants can be costly, making it less economically feasible for some countries or regions.
4. **Nuclear proliferation:** The possibility of the spread of nuclear weapons technology is a major concern associated with the use of nuclear energy.

Future of Nuclear Energy

As concerns about climate change and the need for clean energy sources increase, nuclear energy is likely to play a larger role in meeting the world's energy needs. Some countries, such as France, already rely heavily on nuclear power, while others, such as the United States, are considering expanding their use of nuclear energy.

Advancements in technology may also help to address some of the challenges associated with nuclear energy, such as the development of more advanced reactor designs and improved waste disposal methods.

Conclusion

Nuclear energy is a powerful source of energy that has both advantages and disadvantages. While it provides a reliable, low-carbon source of electricity, it also poses safety and waste management challenges. As we move towards a more sustainable future, it is important to weigh the benefits and risks of nuclear energy alongside other energy sources.

Chapter 34: Green Building Design

Introduction

Green building design is an approach to architecture and construction that emphasizes energy efficiency, resource conservation, and environmental sustainability. The concept of green building design emerged in response to the growing recognition of the negative impact that traditional construction and building practices can have on the environment. The aim of green building design is to create structures that are not only energy-efficient but also minimize the use of natural resources, reduce waste and emissions, and improve indoor air quality. This chapter explores the principles and benefits of green building design.

Principles of Green Building Design

Green building design involves a comprehensive approach to designing and constructing buildings. There are several principles that guide this approach, including:

1. **Energy Efficiency:** One of the primary goals of green building design is to reduce energy consumption. This can be achieved through the use of high-efficiency HVAC systems, energy-efficient lighting, and insulation. In addition, passive solar design can be used to take advantage of natural heating and cooling.

2. **Water Conservation:** Green building design also aims to minimize water consumption. This can be achieved through the use of low-flow fixtures, rainwater harvesting, and greywater reuse systems.

3. **Sustainable Materials:** The use of sustainable materials is another key principle of green building design. This involves using materials that are renewable, recyclable, and non-toxic.
4. **Waste Reduction:** The reduction of waste is an important aspect of green building design. This can be achieved through the use of prefabricated components, recycling, and the use of materials with a longer lifespan.
5. **Indoor Environmental Quality:** Green building design also prioritizes the health and well-being of occupants by improving indoor air quality, acoustics, and access to natural light.

Benefits of Green Building Design: Green building design offers numerous benefits, including:

1. **Energy Savings:** Green buildings use less energy, which can result in significant cost savings over time.
2. **Water Savings:** Green buildings also use less water, reducing water bills and conserving a valuable resource.
3. **Improved Indoor Air Quality:** Green buildings prioritize the health and well-being of occupants by improving indoor air quality and reducing exposure to harmful chemicals.
4. **Reduced Environmental Impact:** Green building design reduces the negative impact that construction and building practices have on the environment.
5. **Increased Property Value:** Green buildings often have a higher property value and can be more

attractive to potential buyers or renters.

Conclusion

Green building design is an important approach to architecture and construction that emphasizes energy efficiency, resource conservation, and environmental sustainability. By following the principles of green building design, we can reduce our impact on the environment and improve the health and well-being of occupants. The benefits of green building design are numerous and far-reaching, from reduced energy and water consumption to improved indoor air quality and increased property value.

Chapter 35: Sustainable Agriculture

Introduction

Agriculture is the backbone of human civilization, providing food and raw materials for the population. However, traditional agricultural practices are becoming unsustainable due to environmental degradation and climate change. Sustainable agriculture is a solution to these problems, and it is based on the principles of environmental, social, and economic sustainability. This chapter will explore sustainable agriculture, its principles, and practices.

What is Sustainable Agriculture?

Sustainable agriculture is an integrated system of practices that enhance soil fertility, promote biodiversity, and reduce pollution. It is based on the principles of ecological balance, social equity, and economic viability. The goal of sustainable agriculture is to provide food security while protecting the environment and ensuring social equity. It involves practices such as organic farming, agroforestry, crop rotation, and integrated pest management.

Principles of Sustainable Agriculture

There are several principles of sustainable agriculture. These principles include:

1. **Enhancing soil fertility:** Sustainable agriculture emphasizes the importance of maintaining soil fertility through practices such as crop rotation, composting, and using organic fertilizers.
2. **Biodiversity:** Sustainable agriculture promotes the use of diverse crop varieties, which increase the

resilience of crops to pests and diseases and reduce the risk of crop failure.

3. **Integrated pest management:** Sustainable agriculture uses pest control methods that are environmentally friendly and reduce the use of chemical pesticides.

4. **Conservation of natural resources:** Sustainable agriculture focuses on the conservation of natural resources such as water, land, and air through practices such as efficient irrigation, conservation tillage, and erosion control.

5. **Social equity:** Sustainable agriculture promotes social equity by ensuring that farmers are paid fairly for their products and that rural communities have access to food.

Practices of Sustainable Agriculture

There are several practices of sustainable agriculture. These practices include:

1. **Organic farming:** Organic farming is a method of agriculture that uses natural methods to enhance soil fertility and control pests and diseases.

2. **Agroforestry:** Agroforestry is the practice of combining trees and crops in the same land area to improve soil fertility and provide multiple products.

3. **Crop rotation:** Crop rotation involves planting different crops in the same field in successive seasons to improve soil fertility and reduce the risk of pests and diseases.

4. **Integrated pest management:** Integrated pest

management involves the use of natural methods such as biological control, crop rotation, and resistant crops to control pests and diseases.

5. **Conservation tillage:** Conservation tillage involves leaving crop residues on the soil surface to reduce soil erosion and improve soil fertility.

Benefits of Sustainable Agriculture

There are several benefits of sustainable agriculture. These benefits include:

1. **Environmental sustainability:** Sustainable agriculture promotes environmental sustainability by reducing the use of chemical pesticides and fertilizers, conserving natural resources, and reducing greenhouse gas emissions.

2. **Social equity:** Sustainable agriculture promotes social equity by ensuring that farmers are paid fairly for their products and that rural communities have access to food.

3. **Economic viability:** Sustainable agriculture provides economic viability by reducing the cost of production, increasing yield, and enhancing soil fertility.

Conclusion

Sustainable agriculture is an important solution to the problems of environmental degradation and climate change. It is based on the principles of environmental, social, and economic sustainability. Sustainable agriculture involves practices such as organic farming, agroforestry, crop rotation, and integrated pest management. It provides several benefits, including environmental sustainability, social equity, and economic viability. Sustainable agriculture is the future of agriculture, and it is important that we embrace it to ensure food security and protect the environment.

Chapter 36: Organic Farming

Organic farming is an agricultural production system that aims to enhance and promote ecological balance, conserve biodiversity, and avoid the use of synthetic inputs such as pesticides and fertilizers. Organic farming practices are based on a set of principles that prioritize soil health, crop rotation, and natural pest management over chemical solutions. The organic approach to farming focuses on sustainable practices that benefit the environment, the farmer, and the consumer.

In this chapter, we will explore the key principles and benefits of organic farming, as well as the challenges and limitations that farmers face when transitioning to organic practices.

Key Principles of Organic Farming

1. **Soil Health:** Organic farmers prioritize soil health as the foundation of their farming practices. They believe that healthy soil is the key to growing healthy crops. Organic farmers use natural fertilizers such as compost, manure, and cover crops to build soil fertility and structure. They also avoid using synthetic fertilizers that can damage soil health in the long term.

2. **Crop Rotation:** Organic farmers practice crop rotation to maintain soil health and reduce the risk of pests and diseases. Crop rotation involves alternating crops in a field over a period of time to avoid planting the same crop repeatedly in the same spot. This practice helps to balance soil nutrients and reduce the

buildup of pests and diseases.

3. **Natural Pest Management:** Organic farmers rely on natural methods to control pests and diseases. They use techniques such as crop rotation, natural predators, and physical barriers to manage pests and diseases without relying on synthetic pesticides.

4. **Biodiversity:** Organic farmers promote biodiversity by growing a variety of crops and preserving natural habitats on their farms. This approach helps to support the ecosystem by providing habitat for beneficial insects, birds, and other wildlife.

Benefits of Organic Farming

1. **Improved Soil Health:** Organic farming practices can improve soil health by promoting the growth of beneficial microorganisms and increasing soil organic matter. This leads to healthier and more resilient soil, which can improve crop yields and reduce the need for synthetic fertilizers.

2. **Reduced Environmental Impact:** Organic farming practices can reduce the environmental impact of agriculture by minimizing the use of synthetic pesticides and fertilizers, which can harm wildlife and pollute waterways. Organic farming practices also help to conserve biodiversity and preserve natural habitats.

3. **Healthier Food:** Organic farming practices can produce food that is free from synthetic pesticides and fertilizers, which can be harmful to human health. Organic foods are also often higher in

nutrients and antioxidants, which can have health benefits.

4. **Economic Benefits:** Organic farming can provide economic benefits to farmers by reducing input costs and increasing crop yields. Organic crops can also fetch a premium price in the market, which can improve the profitability of organic farms.

Challenges of Organic Farming

1. **Transition Period:** Transitioning from conventional farming to organic farming can be a challenging and time-consuming process. The transition period can take up to three years, during which farmers must follow organic practices but cannot yet sell their crops as organic.

2. **Yield Reduction:** Organic farming practices can result in lower crop yields in the short term, as the soil adjusts to the new practices. This can be a financial burden for farmers, who may need to invest in additional labor and equipment to maintain their yields.

3. **Pest and Disease Management:** Organic farmers must rely on natural pest and disease management techniques, which can be less effective than synthetic pesticides. This can lead to lower yields and crop losses in some cases.

4. **Certification Costs:** Organic farmers must pay for certification to sell their crops as organic. This can be a significant cost for small farmers, who may struggle to afford certification fees.

<u>Conclusion</u>

In conclusion, organic farming is a holistic approach to agriculture that prioritizes the health of the soil, the environment, and human health. While transitioning to organic farming can present challenges, such as the transition period and pest management, the benefits to soil health, reduced environmental impact, and healthier food make it a worthwhile endeavor. Organic farming can provide economic benefits to farmers while preserving biodiversity and promoting sustainable farming practices. Overall, organic farming is a promising solution to the challenges facing modern agriculture and can contribute to a healthier and more sustainable future for our planet.

Chapter 37: Aquaponics

Aquaponics is a sustainable farming method that combines aquaculture and hydroponics in a closed-loop system. It is a type of aquaculture that involves the cultivation of aquatic animals and plants in a symbiotic environment. Aquaponics is a sustainable way to grow crops and fish without the use of pesticides, fertilizers, or other harmful chemicals.

The concept of aquaponics dates back to the ancient Aztecs, who used a similar method to cultivate crops in floating gardens called chinampas. Today, aquaponics is gaining popularity as a sustainable farming method due to its many advantages over traditional agriculture.

In an aquaponic system, fish and plants are raised together in a closed-loop system. The fish produce waste, which is rich in nitrogen and other nutrients that the plants need to grow. The plants, in turn, absorb these nutrients and filter the water, removing harmful waste products that could harm the fish.

The basic components of an aquaponic system include a fish tank, a grow bed, and a water pump. The fish tank is where the fish are raised and the grow bed is where the plants are grown. The water pump circulates water between the two components, delivering nutrients to the plants and oxygen to the fish.

There are several different types of aquaponic systems, including deep water culture, nutrient film technique, and media-based systems. Deep water culture is the simplest type of aquaponic system and involves suspending the plants in nutrient-rich water. Nutrient film technique involves growing plants in channels filled with nutrient-rich water that is constantly flowing. Media-based systems use a growing medium, such as gravel or clay, to support the plants and provide a surface for bacteria to grow.

One of the advantages of aquaponics is that it is a highly efficient method of farming. It uses up to 90% less water than traditional farming methods, as the water is recycled and reused within the closed-loop system. Additionally, aquaponics can produce more food per square foot than traditional farming methods, making it an ideal method for urban farming.

Aquaponics is also a sustainable farming method, as it does not rely on harmful chemicals or fertilizers that can harm the environment. It is a natural way to grow crops and fish, which is beneficial for both the environment and human health.

In addition to being sustainable, aquaponics can also be a profitable business venture. It can be used to grow a wide variety of crops and fish, including vegetables, herbs, fruits, and even ornamental plants. Aquaponics can also be used to grow high-value crops, such as gourmet mushrooms, microgreens, and specialty herbs.

However, aquaponics does have some limitations. It requires an initial investment to set up the system and may require ongoing maintenance to ensure that the system remains healthy. Additionally, not all crops are suitable for aquaponic farming, as some may require specific growing conditions that cannot be met in an aquaponic system.

Overall, aquaponics is a promising sustainable farming method that has many advantages over traditional agriculture. It offers a natural, chemical-free way to grow crops and fish, and can be used to produce a wide variety of high-value crops. With proper planning and maintenance, aquaponics can be a profitable and sustainable farming method for farmers and urban growers alike.

Chapter 38: Recycling

Introduction

Recycling is the process of converting waste materials into new materials and objects, thereby reducing the consumption of new raw materials and energy usage. Recycling has numerous environmental, economic, and social benefits. In this chapter, we will discuss the concept of recycling, its importance, the recycling process, and the different materials that can be recycled.

The Importance of Recycling

Recycling is essential to reduce the negative impact of waste on the environment. By recycling, we can conserve natural resources and reduce pollution. Recycling helps in reducing greenhouse gas emissions, as it requires less energy to recycle materials than to produce new ones. It also reduces the amount of waste that ends up in landfills and reduces the need for new landfills, saving land resources. Moreover, recycling creates job opportunities and promotes sustainable economic growth.

The Recycling Process The recycling process involves several steps, which are as follows:

1. **Collection:** The first step in the recycling process is the collection of recyclable materials. Different methods are used for collecting recyclable materials, including curbside collection, drop-off centers, and recycling bins.
2. **Sorting:** After the collection, the recyclable materials are sorted based on their type and quality. The

sorting process involves separating materials such as plastic, paper, glass, and metal.

3. **Cleaning:** Once sorted, the materials are cleaned to remove any impurities such as dirt or food residue.

4. **Processing:** After cleaning, the materials are processed to be reused. The processing can involve melting, shredding, or grinding the materials to create raw materials for manufacturing.

5. **Manufacturing:** The final step in the recycling process is the manufacturing of new products from the processed materials.

Materials that can be Recycled Various materials can be recycled, including:

1. **Paper:** Paper can be recycled into new paper products such as newspapers, tissue paper, and cardboard boxes.

2. **Plastic:** Plastic can be recycled into various products such as plastic bottles, containers, and packaging materials.

3. **Glass:** Glass can be recycled into new glass products such as jars, bottles, and glassware.

4. **Metal:** Metal can be recycled into new metal products such as cans, bicycles, and car parts.

5. **Electronics:** Electronic waste such as computers, televisions, and cell phones can be recycled to recover valuable metals such as gold and copper.

Challenges in Recycling Despite its benefits, recycling faces various challenges, including:

1. **Contamination:** Contamination is a significant problem in recycling as it can affect the quality of recycled materials. Contamination can occur due to the presence of non-recyclable materials or improper handling of recyclables.
2. **Lack of Awareness:** Many people are not aware of the benefits of recycling and do not understand how to recycle properly.
3. **Infrastructure:** Recycling requires proper infrastructure such as recycling centers, collection facilities, and processing plants. Lack of infrastructure can hinder the recycling process.
4. **Cost:** Recycling can be more expensive than producing new materials due to the cost of collection, transportation, sorting, and processing.

Conclusion

Recycling is a crucial aspect of environmental conservation and sustainable development. By reducing waste and conserving natural resources, recycling helps to mitigate climate change and promote sustainable economic growth. It is important to increase awareness about the benefits of recycling and invest in infrastructure to enhance the recycling process.

Chapter 39: Composting

Composting is the process of converting organic waste into a rich, soil-like substance called compost. This substance is an excellent fertilizer and soil amendment, and it helps to reduce the amount of waste sent to landfills. Composting is a simple, natural process that can be done in your own backyard, and it has many benefits for both the environment and your garden.

The composting process involves a combination of organic waste materials, such as food scraps, yard trimmings, and other natural materials, and microorganisms such as bacteria and fungi that break down the materials into a nutrient-rich soil amendment. The process takes a few months to complete and requires a combination of organic matter, oxygen, moisture, and the right temperature.

There are different methods of composting, each with its own advantages and disadvantages. One of the most popular methods is the traditional method, which involves creating a pile of organic matter and turning it regularly to aerate it. This method can take several months to complete, but it produces a high-quality compost that is rich in nutrients and beneficial microorganisms.

Another popular method is vermicomposting, which involves using earthworms to break down the organic matter. This method is faster than traditional composting and produces a high-quality compost that is rich in nutrients and beneficial microorganisms.

Composting has many benefits for the environment. By composting organic waste, you can reduce the amount of waste sent to landfills, which in turn reduces the amount of methane gas that is produced. Methane is a potent greenhouse gas that contributes to climate change. Composting also helps to reduce the need for chemical fertilizers, which can pollute groundwater and harm wildlife.

Composting has many benefits for your garden as well. Compost improves soil structure, which allows for better drainage and air circulation. It also adds nutrients to the soil, which can improve plant growth and health. Compost also helps to retain moisture in the soil, reducing the need for irrigation.

To start composting at home, you need a compost bin or pile, organic matter, and a few simple tools. The organic matter can be a combination of food scraps, yard waste, and other natural materials. The tools you need include a pitchfork or shovel to turn the compost pile, a hose or watering can to add moisture, and a thermometer to monitor the temperature of the compost pile.

When composting, it's important to remember a few things. First, the compost pile should be located in a spot that is well-drained and receives plenty of sunlight. Second, the pile should be turned regularly to allow air to circulate and speed up the composting process. Third, the pile should be kept moist but not waterlogged. Finally, the pile should be monitored for temperature to ensure that it is within the optimal range for composting.

In conclusion, composting is a simple and effective way to reduce waste, improve soil health, and promote sustainable gardening practices. By composting at home, you can reduce your environmental footprint, save money on fertilizer, and create a healthy, vibrant garden.

Chapter 40: Waste Reduction

Introduction

Waste reduction is an essential aspect of environmental sustainability, and it is the process of reducing the amount of waste generated by individuals, businesses, and communities. It involves using resources efficiently and minimizing waste by reducing, reusing, and recycling materials.

Waste is generated from a variety of sources, including households, industries, and businesses. This waste can take different forms, such as organic waste, hazardous waste, electronic waste, and plastic waste. Each type of waste requires a unique approach to reduce, reuse, and recycle it.

This chapter explores waste reduction as an important part of environmental sustainability, discussing the benefits of waste reduction, approaches to waste reduction, and waste reduction strategies that can be adopted by individuals and organizations.

Benefits of Waste Reduction

Waste reduction has numerous benefits for the environment, economy, and society. Some of these benefits include:

1. Environmental Benefits

Waste reduction reduces the amount of waste that ends up in landfills, reducing the environmental impact of waste. This can prevent soil and water contamination and reduce greenhouse gas emissions that result from the decomposition of waste in landfills.

1. Economic Benefits

Waste reduction can result in significant cost savings for individuals, businesses, and communities. It reduces the cost of waste disposal, saves resources and energy, and creates jobs in recycling and waste reduction industries.

1. Social Benefits

Waste reduction promotes social responsibility and encourages individuals and communities to take responsibility for their waste. It also encourages the adoption of sustainable practices that can reduce environmental impact and create a healthier living environment.

Approaches to Waste Reduction

There are three primary approaches to waste reduction: reduce, reuse, and recycle. Each approach focuses on reducing waste in different ways.

1. Reduce

Reducing waste involves minimizing the amount of waste generated by individuals, businesses, and communities. This can be achieved through the following strategies:

- **Avoiding single-use products:** Single-use products, such as plastic bags and water bottles, generate a significant amount of waste. Avoiding their use can reduce waste.

- **Choosing durable and long-lasting products:** Durable and long-lasting products reduce waste by reducing the need for frequent replacements.

- **Reducing packaging:** Packaging generates a significant amount of waste. Choosing products with minimal packaging or using reusable containers can reduce waste.

- **Minimizing food waste:** Food waste generates a significant amount of waste. Minimizing food waste by reducing portion sizes, composting food scraps, and donating excess food can reduce waste.

1. Reuse

Reusing waste involves finding alternative uses for waste materials to reduce the amount of waste generated. This can be achieved through the following strategies:

- **Repairing products:** Repairing products rather than replacing them can extend their lifespan and reduce waste.

- **Donating or selling items:** Donating or selling items that are no longer needed or wanted can reduce waste by providing others with useful items.

- **Using reusable products:** Using reusable products, such as cloth bags and water bottles, can reduce waste by eliminating the need for single-use products.

1. Recycle

Recycling waste involves collecting and processing waste materials to produce new products. This can be achieved through the following strategies:

- **Separating recyclable materials:** Separating recyclable materials, such as paper, plastic, and metal, from other waste materials makes them easier to recycle.

- **Participating in community recycling programs:** Participating in community recycling programs, such as curbside recycling, makes it easier to recycle waste.

- **Choosing products made from recycled materials:** Choosing products made from recycled materials supports the recycling industry and reduces the demand for new materials.

Waste Reduction Strategies

Waste reduction is the process of decreasing the amount of waste generated, with the aim of minimizing the environmental impact of waste. It involves several strategies and practices, such as reducing, reusing, and recycling waste. This chapter will explore various waste reduction strategies and how they can be applied in our daily lives.

1. Reduce

Reducing waste is the most effective way to minimize waste production. This involves limiting the consumption of products and services that generate waste. Some ways to reduce waste include:

- **Purchasing only what you need:** Avoid impulse purchases and only buy what you need.

- Using reusable bags: Bring your own reusable bags when shopping.

- **Using reusable containers:** Avoid using single-use plastic containers and use reusable containers instead.

- **Using a refillable water bottle:** Instead of buying bottled water, use a refillable water bottle.

- **Printing less:** Print only what you need and avoid printing unnecessary documents.

1. Reuse

Reuse involves finding ways to use products and materials again, rather than disposing of them after a single use. Some ways to reuse products include:

- **Using cloth napkins:** Use cloth napkins instead of disposable paper napkins.

- **Using reusable cups and plates:** Use reusable cups and plates instead of disposable paper cups and plates.

- **Donating items:** Donate unwanted items such as clothing, furniture, and electronics to charity or a thrift store.

- **Refilling ink cartridges:** Refill ink cartridges instead of buying new ones.

1. Recycle

Recycling involves converting waste materials into new products. Recycling reduces the need to extract new raw materials from the environment, thereby conserving natural resources. Some ways to recycle include:

- **Sorting waste:** Sort waste into different categories such as paper, plastic, and glass.

- **Recycling electronics:** Recycle electronics such as cell phones and computers.

- **Recycling household items:** Recycle household items such as cardboard, paper, and plastic bottles.

1. Composting

Composting involves the decomposition of organic waste into a nutrient-rich soil amendment. Composting is an effective way to reduce the amount of waste that ends up in landfills. Some ways to compost include:

- **Composting food scraps:** Compost food scraps such as fruit and vegetable peels, eggshells, and coffee grounds.

- **Composting yard waste:** Compost yard waste such as leaves, grass clippings, and tree branches.

1. Waste Reduction at Work

Waste reduction strategies can also be applied in the workplace. Some ways to reduce waste at work include:

- **Using digital documents:** Use digital documents instead of printing out paper documents.

- **Recycling:** Provide recycling bins in the workplace.

- **Providing reusable items:** Provide reusable items such as cups, plates, and utensils in the workplace.

Conclusion

Waste reduction strategies are essential in minimizing the environmental impact of waste. The implementation of these strategies can help conserve natural resources, reduce greenhouse gas emissions, and minimize the amount of waste generated. By reducing, reusing, and recycling waste, we can create a more sustainable future for ourselves and future generations.

Chapter 41: Environmental Laws and Regulations

Introduction

Environmental laws and regulations are in place to protect the environment and the health of the human population. These laws are designed to limit pollution, regulate hazardous materials, and protect natural resources. This chapter will discuss the history of environmental regulations, their purpose, and the different types of laws and regulations that are currently in place.

The History of Environmental Regulations

Environmental regulations have been in place for many years, but the first major federal law did not come into effect until the 1970s. The Clean Air Act of 1970 was the first major environmental law in the United States. It was designed to regulate air pollution and limit the release of hazardous pollutants into the atmosphere. This law was followed by the Clean Water Act of 1972, which regulated the discharge of pollutants into water sources.

Purpose of Environmental Regulations

The purpose of environmental regulations is to protect the environment and human health by limiting the impact of human activities on the natural world. These regulations aim to reduce pollution, prevent the degradation of natural resources, and promote sustainable practices. Environmental regulations can also be used to encourage businesses to adopt more sustainable practices, such as reducing waste and increasing energy efficiency.

Types of Environmental Regulations

There are many different types of environmental regulations in place, including laws and regulations that target air quality, water quality, hazardous waste, and other environmental concerns. Some of the most important environmental regulations include:

1. **The Clean Air Act:** This law regulates the emissions of air pollutants from various sources, including factories, power plants, and transportation.
2. **The Clean Water Act:** This law regulates the discharge of pollutants into water sources and the quality of water in streams, rivers, and lakes.
3. **The Resource Conservation and Recovery Act:** This law regulates the management and disposal of hazardous waste, including the storage, transportation, and treatment of these materials.
4. **The Endangered Species Act:** This law protects endangered and threatened species by regulating the use of their habitats and preventing the harm or destruction of these habitats.
5. **The National Environmental Policy Act:** This law requires federal agencies to evaluate the environmental impact of any major federal action, such as building a new highway or constructing a new dam.

Waste Reduction Strategies

Waste reduction strategies are an important part of environmental regulations and can help to reduce the impact of human activities on the natural world. Some of the most effective waste reduction strategies include:

1. **Recycling:** Recycling is the process of converting waste materials into new products. Recycling can help to reduce the amount of waste that ends up in landfills, conserve natural resources, and reduce greenhouse gas emissions.

2. **Composting:** Composting is the process of breaking down organic materials, such as food waste and yard waste, into a nutrient-rich soil amendment. Composting can help to reduce the amount of waste that ends up in landfills and can also improve soil health.

3. **Waste Reduction:** Waste reduction strategies include reducing the amount of waste generated in the first place by using durable and reusable products, reducing packaging, and adopting more sustainable practices.

Conclusion

Environmental regulations are in place to protect the environment and human health by limiting the impact of human activities on the natural world. These regulations are designed to reduce pollution, prevent the degradation of natural resources, and promote sustainable practices. Waste reduction strategies are an important part of environmental regulations and can help to reduce the impact of human activities on the natural world. By working together to protect the environment, we can create a healthier and more sustainable future for ourselves and for future generations.

Chapter 42: Environmental Policy

Environmental policy refers to a set of principles and guidelines that govern human activities that affect the natural environment. Environmental policy is designed to protect the environment and promote sustainable practices, while also balancing economic and social considerations. Environmental policy is implemented at the international, national, regional, and local levels.

Environmental policy is a complex and multifaceted issue. It involves a wide range of stakeholders, including government agencies, non-governmental organizations (NGOs), businesses, and communities. Environmental policies can be proactive, reactive, or both. Proactive policies are aimed at preventing environmental problems before they occur, while reactive policies are aimed at addressing existing environmental problems.

There are a number of different types of environmental policy, including regulatory policies, economic incentives, and voluntary programs. Regulatory policies are typically the most effective at achieving environmental goals, as they set specific standards and requirements that must be met by businesses and individuals. Economic incentives, such as taxes and subsidies, can also be effective in promoting sustainable practices. Voluntary programs, such as certification programs and labeling schemes, can be useful in promoting sustainability, but they are often less effective than regulatory policies and economic incentives.

Environmental policy can also be classified based on the level of government that is responsible for implementing it. International environmental policy is designed to address global environmental issues, such as climate change, biodiversity loss, and pollution. National environmental policy is developed and implemented by national governments, and it typically focuses on issues that are specific to that country. Regional and local environmental policy is designed to address issues that are specific to a particular region or locality.

Environmental policy has become increasingly important in recent years, as environmental problems such as climate change, pollution, and biodiversity loss have become more pressing. Governments and other stakeholders have recognized the need to develop policies that address these issues and promote sustainable practices. However, the implementation of environmental policies can be challenging, as it often involves balancing competing interests and priorities.

In order to be effective, environmental policy must be based on sound scientific principles and must take into account the social and economic implications of environmental decisions. It must also be flexible enough to adapt to changing circumstances and new information. Finally, environmental policy must be enforced in a fair and consistent manner, in order to ensure that everyone is held accountable for their actions.

In conclusion, environmental policy is a critical component of efforts to protect the environment and promote sustainable practices. It is a complex and multifaceted issue that involves a wide range of stakeholders and requires a careful balancing of economic, social, and environmental considerations. Environmental policy must be based on sound scientific principles, must be flexible and adaptive, and must be enforced in a fair and consistent manner. Only then can we hope to achieve the goal of a sustainable future for ourselves and for future generations.

Chapter 43: Environmental Ethics

Introduction

Environmental ethics is a branch of philosophy that examines the moral and ethical issues related to the environment and its resources. It considers the relationship between humans and the natural world and seeks to determine how humans ought to behave towards the environment. Environmental ethics is an important field of study because it can help individuals and societies understand their responsibilities to the environment and guide their actions to protect it.

Historical Development

Environmental ethics has its roots in ancient philosophies such as Taoism, which emphasized the interconnectedness of all things in nature, and Stoicism, which saw humans as part of a larger natural system. However, the modern environmental movement and the increasing awareness of environmental issues in the 20th century led to the development of environmental ethics as a distinct field of study.

Principles of Environmental Ethics

There are several principles of environmental ethics that can guide individuals and societies in their interactions with the environment. Some of these principles include:

1. **Intrinsic Value:** This principle asserts that the environment has inherent value and should be protected for its own sake, not just for its usefulness to humans.

2. **Ecocentrism:** This principle recognizes that humans are not the only beings with inherent value, and that

the well-being of ecosystems and the natural world as a whole should be a primary concern.

3. **Anthropocentrism:** This principle sees humans as the most important beings in the natural world, and that the environment should be protected only to the extent that it benefits humans.

4. **Biocentrism:** This principle holds that all living beings have inherent value and should be protected.

5. **Sustainability:** This principle emphasizes the need to use resources in a way that preserves them for future generations.

Environmental Ethics in Practice

Environmental ethics has real-world applications in a variety of contexts, from environmental activism to policy-making. For example, environmental activists may use the principle of intrinsic value to argue for the protection of endangered species, while policymakers may use the principle of sustainability to guide decisions about resource management.

Environmental ethics can also be applied in individual decision-making. For example, an individual may choose to reduce their carbon footprint by biking instead of driving, or by using energy-efficient appliances in their home. These choices reflect a concern for the environment and a desire to act in accordance with environmental ethics.

Conclusion

Environmental ethics is an important field of study that seeks to promote responsible behavior towards the environment. Its principles can guide individuals and societies in their interactions with the natural world, and can help us create a more sustainable and just future. By understanding and applying environmental ethics, we can work towards a healthier, more equitable planet for all beings.

Chapter 44: Environmental Education

Introduction

Environmental education is the process of imparting knowledge and skills to individuals about the environment and how it works. The goal of environmental education is to raise awareness about environmental issues, develop critical thinking skills, and encourage individuals to take action to address environmental problems. In this chapter, we will explore the importance of environmental education, the different approaches to teaching environmental education, and the challenges and opportunities that exist in the field of environmental education.

The Importance of Environmental Education

Environmental education is critical for several reasons. First, it raises awareness about environmental issues and helps individuals understand the impacts of human activities on the environment. Second, it provides individuals with the knowledge and skills they need to make informed decisions about their actions and their impact on the environment. Third, it promotes critical thinking skills, enabling individuals to analyze complex environmental problems and develop creative solutions. Finally, environmental education can inspire individuals to take action and make a positive impact on the environment.

Approaches to Teaching Environmental Education There are several approaches to teaching environmental education, including:

1. **Place-based education:** This approach focuses on using the local environment as a context for learning. By studying the local environment, individuals can gain a deeper understanding of the ecological, social, and economic systems that are present in their community.

2. **Experiential education:** This approach involves hands-on learning experiences that enable individuals to connect with the environment and develop a sense of place. Experiential education can include outdoor activities, field trips, and service learning projects.

3. **Inquiry-based learning:** This approach involves asking questions and conducting investigations to explore environmental issues. By using scientific inquiry, individuals can develop critical thinking skills and gain a deeper understanding of environmental issues.

4. **Multidisciplinary education:** This approach involves integrating environmental education into other subject areas, such as science, social studies, and language arts. By connecting environmental education to other subject areas, individuals can develop a more holistic understanding of the environment and its importance.

Challenges and Opportunities in Environmental Education

Despite the importance of environmental education, there are several challenges and opportunities that exist in the field. One of the biggest challenges is the lack of funding and resources for environmental education programs. Many schools and organizations struggle to provide adequate resources for environmental education, which can limit the impact of these programs.

Another challenge is the need to develop effective strategies for engaging diverse communities in environmental education. Environmental issues can impact different communities in different ways, and it is important to ensure that all communities have access to environmental education programs that meet their specific needs.

Finally, there are several opportunities in the field of environmental education. For example, advances in technology have made it possible to develop new and innovative ways of teaching environmental education, such as online learning platforms and virtual reality experiences. Additionally, there is growing recognition of the importance of environmental education in preparing individuals for the challenges of the 21st century, including climate change, resource depletion, and environmental degradation.

Conclusion

Environmental education is critical for raising awareness about environmental issues, developing critical thinking skills, and encouraging individuals to take action to address environmental problems. By using a variety of approaches to teaching environmental education and addressing the challenges and opportunities in the field, we can ensure that individuals have the knowledge and skills they need to make informed decisions and take positive action to protect the environment.

Chapter 45: Sustainable Transportation

Transportation is an essential component of modern society and enables the movement of people, goods, and services. However, transportation also has significant impacts on the environment, including air and noise pollution, greenhouse gas emissions, and habitat destruction. Therefore, it is important to develop sustainable transportation systems that are environmentally friendly, efficient, and safe. This chapter will discuss the concept of sustainable transportation, the challenges of achieving it, and the various strategies and technologies that can help create a more sustainable transportation system.

What is Sustainable Transportation?

Sustainable transportation is a concept that seeks to provide safe, efficient, and affordable transportation while minimizing the negative impacts on the environment and society. The goal is to create a transportation system that meets the needs of present and future generations without compromising the ability of future generations to meet their own needs.

Challenges of Achieving Sustainable Transportation

The primary challenge of achieving sustainable transportation is the dependence on fossil fuels. Fossil fuels such as gasoline and diesel power most of the transportation systems around the world, which leads to significant greenhouse gas emissions and air pollution. Additionally, the limited supply of fossil fuels and the geopolitical instability surrounding them can cause price volatility and supply disruptions. Developing and adopting alternative fuels and modes of transportation that are environmentally friendly, efficient, and cost-effective is essential to achieving sustainable transportation.

Another challenge is the lack of infrastructure and policies to support sustainable transportation. For example, many cities lack the proper infrastructure for bike lanes, public transportation, and electric vehicle charging stations, which makes it difficult for people to use these modes of transportation. Additionally, policies such as tax incentives and regulations can encourage the adoption of sustainable transportation practices.

Strategies and Technologies for Sustainable Transportation

1. Alternative Fuels

Alternative fuels are fuels that are not derived from fossil fuels. They include biofuels, such as ethanol and biodiesel, and renewable electricity, such as solar, wind, and hydropower. These fuels emit fewer greenhouse gases than fossil fuels, making them more environmentally friendly. Additionally, many alternative fuels can be produced domestically, reducing dependence on foreign oil and improving energy security.

1. Electric Vehicles

Electric vehicles (EVs) are powered by an electric motor that uses electricity from batteries or fuel cells, rather than gasoline or diesel. EVs emit fewer greenhouse gasses and air pollutants than gasoline-powered vehicles and are becoming more affordable and practical for everyday use. Additionally, the use of renewable electricity to power EVs can make them even more environmentally friendly.

1. Public Transportation

Public transportation, such as buses and trains, can be a more environmentally friendly alternative to individual cars. Public transportation systems can reduce traffic congestion, air pollution, and greenhouse gas emissions. Additionally, public transportation can be a more affordable option for people who cannot afford to own a car.

1. Active Transportation

Active transportation includes walking, biking, and other non-motorized modes of transportation. Active transportation can improve physical health, reduce traffic congestion, and reduce air pollution. However, it requires proper infrastructure and policies to support its use, such as bike lanes and pedestrian walkways.

1. Transportation Demand Management

Transportation demand management (TDM) strategies aim to reduce the demand for transportation by encouraging alternatives to driving, such as carpooling, telecommuting, and flexible work schedules. TDM can reduce traffic congestion, air pollution, and greenhouse gas emissions, while also saving time and money for commuters.

1. Smart Growth

Smart growth is a land-use planning strategy that aims to create compact, walkable communities that are designed to reduce the need for driving. Smart growth can reduce traffic congestion, air pollution, and greenhouse gas emissions, while also promoting economic development and social equity.

1. Green Infrastructure

Green infrastructure refers to natural and semi-natural features that provide ecological, social, and economic benefits. It is a term used to describe the use of natural systems and processes in urban and suburban environments to provide environmental and economic benefits. Green infrastructure includes trees, wetlands, parks, gardens, green roofs, permeable pavements, and other natural features. It can help to reduce the negative impacts of human activities on the environment by improving air and water quality, reducing urban heat islands, and providing habitat for wildlife.

Green infrastructure can also provide numerous benefits for human health and wellbeing. Trees and other vegetation can help to reduce air pollution and the urban heat island effect, which can lead to reduced rates of respiratory disease and heat-related illness. Access to green spaces can also help to reduce stress and improve mental health, and green infrastructure projects can create opportunities for recreation and social interaction.

One example of a green infrastructure project is the use of green roofs, which involve the installation of vegetation on rooftops of buildings. Green roofs can help to reduce energy use by providing insulation, and they can also help to reduce stormwater runoff by absorbing and filtering rainwater. Another example is the use of permeable pavements, which allow rainwater to seep through the pavement and into the ground, reducing runoff and improving water quality.

Sustainable Transportation

Sustainable transportation refers to the use of transportation methods and systems that are environmentally friendly and economically sustainable. This includes modes of transportation such as walking, biking, and public transit, as well as the use of low-emission vehicles and alternative fuels.

One of the main goals of sustainable transportation is to reduce the negative impacts of transportation on the environment, including air pollution, greenhouse gas emissions, and habitat destruction. Sustainable transportation can also help to reduce traffic congestion and improve public health by promoting physical activity and reducing noise pollution.

One example of sustainable transportation is the use of public transit systems such as buses, trains, and light rail. Public transit can help to reduce the number of cars on the road, which can reduce traffic congestion and air pollution. Another example is the use of electric vehicles, which produce fewer emissions than traditional gasoline-powered vehicles.

In addition to promoting sustainable modes of transportation, there are also a variety of strategies that can be used to reduce the environmental impact of transportation. These include improving vehicle fuel efficiency, promoting carpooling and ridesharing, and developing infrastructure for alternative modes of transportation such as bike lanes and pedestrian walkways.

Conclusion

Environmental sustainability is a complex issue that requires a multidisciplinary approach. It involves the development of policies and strategies that promote the protection and conservation of natural resources, as well as the implementation of technologies and practices that reduce the negative impacts of human activities on the environment.

Sustainability efforts must be integrated into all aspects of society, including transportation, agriculture, energy production, and waste management. It is also important to consider the social and economic impacts of sustainability initiatives, and to ensure that they are accessible and equitable for all members of society.

By working together to promote environmental sustainability, we can help to protect our planet and ensure a healthy and prosperous future for generations to come.

Chapter 46: The Future of Earth and Environmental Science

As our understanding of the Earth and its systems continues to evolve, the future of earth and environmental science holds a great deal of promise. As technology advances and more data becomes available, we can expect to gain even deeper insights into the complex interactions that drive our planet's functioning. Here are some of the key areas of research and development that we can expect to see in the coming years:

1. **Climate Modeling and Prediction:** As climate change continues to impact the planet, researchers are working to improve our ability to predict future changes in the Earth's climate. This work includes developing more sophisticated climate models, better understanding the impacts of feedback loops, and identifying potential tipping points beyond which the climate may undergo rapid and irreversible changes.

2. **Remote Sensing:** The use of remote sensing technologies such as satellite imagery, LiDAR, and unmanned aerial vehicles (UAVs) is rapidly expanding our ability to monitor and understand the Earth's systems. As these technologies become more affordable and accessible, we can expect to see increased use in environmental monitoring, disaster response, and conservation efforts.

3. **Biotechnology:** Advances in biotechnology are opening up new avenues for understanding and

managing the Earth's ecosystems. Gene editing and synthetic biology offer the potential to modify organisms to better adapt to changing environmental conditions or to restore damaged ecosystems.

4. **Big Data:** The explosion of data from remote sensing, social media, and other sources is creating new opportunities for earth and environmental science research. By harnessing the power of big data analytics, researchers can gain insights into complex patterns and relationships that were previously hidden.

5. **Sustainability Science:** As humanity continues to put increasing pressure on the Earth's resources, sustainability science is becoming an increasingly important field of study. This includes developing new approaches to resource management, reducing waste and emissions, and finding ways to build more sustainable communities.

6. **Earth Observation Systems:** Advances in Earth observation systems are making it possible to continuously monitor the Earth's systems at unprecedented levels of detail. This includes the development of new satellites, airborne sensors, and ground-based observatories that can capture data on everything from ocean currents to atmospheric chemistry.

7. **Environmental Policy:** As we continue to grapple with the complex environmental challenges facing the planet, environmental policy will play an increasingly important role in shaping the future of

earth and environmental science. This includes developing new regulations, incentives, and strategies to promote sustainability and protect the environment.

Overall, the future of earth and environmental science is bright, as researchers, policymakers, and citizens work together to understand and manage the complex systems that make up our planet. By embracing new technologies, data-driven approaches, and innovative policy solutions, we can build a more sustainable and resilient future for ourselves and for generations to come.

Chapter 47: The Role of Technology in Environmental Protection

Technology plays a significant role in environmental protection, from developing clean energy sources to creating more efficient manufacturing processes. In recent years, technology has been advancing at an unprecedented rate, and this has opened up new opportunities for solving environmental challenges. In this chapter, we will explore the various ways in which technology is being used to protect the environment.

Clean Energy Technologies

Clean energy technologies are designed to reduce the reliance on fossil fuels, which are a major source of greenhouse gas emissions. Renewable energy sources such as solar, wind, hydroelectric, and geothermal power are becoming increasingly popular as a means of generating electricity. These technologies have a much lower impact on the environment compared to traditional energy sources, and they are becoming more affordable as technology advances.

Solar Energy

Solar energy is one of the most widely used renewable energy sources, and it has the potential to power the entire planet. Solar panels are becoming increasingly efficient, and new technologies are being developed to make solar energy more affordable and accessible to everyone. Solar farms are being built all over the world, and many homes and businesses are now installing solar panels to reduce their energy costs and carbon footprint.

Wind Energy

Wind energy is another widely used renewable energy source that is becoming increasingly popular. Wind turbines are being built all over the world, and they are capable of generating large amounts of electricity. Advances in technology are making wind turbines more efficient, and new designs are being developed to make them more aesthetically pleasing and less disruptive to wildlife.

Hydroelectric Energy

Hydroelectric power is generated by harnessing the power of moving water. Dams are built to create a reservoir of water, which is then used to power turbines and generate electricity. Hydroelectric power is a clean and reliable source of energy, and it has been used for decades to power homes and businesses around the world.

Geothermal Energy

Geothermal energy is generated by harnessing the natural heat of the earth. Heat pumps are used to extract heat from underground and transfer it to buildings, where it is used for heating and cooling. Geothermal energy is a clean and renewable source of energy, and it has the potential to power millions of homes and businesses around the world.

Energy Efficiency Technologies

In addition to clean energy technologies, energy efficiency technologies are also being developed to reduce energy consumption and greenhouse gas emissions. Energy-efficient buildings, appliances, and transportation are all part of the solution to the environmental challenges we face.

Green Buildings

Green buildings are designed to be energy-efficient and environmentally friendly. They are built with materials that are sustainable and non-toxic, and they are designed to maximize natural light and ventilation. Green buildings use less energy than traditional buildings, and they can also produce their own energy through the use of solar panels and other renewable energy sources.

Smart Grids

Smart grids are being developed to make the electricity grid more efficient and reliable. They use advanced sensors and communication technologies to monitor and control the flow of electricity, and they can also integrate renewable energy sources such as solar and wind power. Smart grids can reduce energy consumption and greenhouse gas emissions by making the electricity grid more efficient and responsive to changing demand.

Electric Vehicles

Electric vehicles are becoming increasingly popular as a means of reducing greenhouse gas emissions from transportation. Advances in battery technology have made electric vehicles more affordable and practical, and new charging infrastructure is being built to support their widespread adoption. Electric vehicles have the potential to revolutionize transportation, and they could play a major role in reducing greenhouse gas emissions from the transportation sector.

Conclusion

In conclusion, technology has an essential role in environmental protection. Clean energy technologies such as solar, wind, hydroelectric, and geothermal power are increasingly becoming popular as a means of reducing greenhouse gas emissions and providing clean and renewable energy. Energy efficiency technologies like green buildings, smart grids, and electric vehicles are also contributing to reducing energy consumption and greenhouse gas emissions. As technology continues to advance, it will provide new solutions to the environmental challenges we face, and it is crucial that we continue to invest in these technologies to protect our planet for future generations.

Chapter 48: The Effects of Climate Change on Agriculture

Agriculture is one of the most important sectors in the global economy. It plays a vital role in providing food, fiber, and fuel for the growing population. However, climate change is posing a significant threat to agriculture, with its impact being felt in various parts of the world. The changing climate is causing shifts in rainfall patterns, rising temperatures, and extreme weather events, which are having adverse effects on agricultural productivity and food security.

Shifts in rainfall patterns Climate change is causing shifts in rainfall patterns, which are affecting agricultural production. Some areas are experiencing increased rainfall, leading to flooding and soil erosion, while others are experiencing decreased rainfall, causing drought and desertification. These shifts are making it difficult for farmers to plan their planting schedules and manage their crops effectively.

Rising temperatures The Earth's temperature is rising, leading to heat stress in plants and animals. High temperatures can cause wilting, reduced photosynthesis, and reduced yields in crops. It can also cause livestock to become stressed, leading to reduced milk production, weight loss, and even death. Furthermore, rising temperatures can increase the incidence of pests and diseases, which can further reduce crop yields.

Extreme weather events Climate change is causing an increase in the frequency and intensity of extreme weather events, such as floods, droughts, hurricanes, and tornadoes.

These events are affecting agricultural production, with floods destroying crops and droughts causing crop failure. Hurricanes and tornadoes are also causing damage to agricultural infrastructure, such as barns, storage facilities, and irrigation systems.

Food security Climate change is affecting food security by reducing agricultural productivity and increasing food prices. With a growing population, it is becoming increasingly difficult to meet the demand for food. The impacts of climate change on agriculture are making it even more challenging to ensure food security for all.

Adaptation strategies To address the challenges posed by climate change on agriculture, there are several adaptation strategies that can be implemented. These include:

1. **Crop diversification:** Planting a variety of crops can help reduce the risk of crop failure due to extreme weather events.

2. **Irrigation:** Investing in irrigation systems can help ensure that crops have access to water during droughts.

3. **Improved soil management:** Practices such as conservation tillage, cover cropping, and crop rotation can help improve soil health, making crops more resilient to climate change.

4. **Agroforestry:** Planting trees on farms can help reduce soil erosion, improve soil health, and provide shade for crops.

5. **Improved livestock management:** Providing shade, water, and good nutrition to livestock can help

reduce the impact of heat stress on animals.

6. **Early warning systems:** Investing in early warning systems can help farmers plan for extreme weather events and take action to reduce their impact.

7. **Research and development:** Investing in research and development can help identify new crops and management practices that are better suited to the changing climate.

Conclusion

Climate change is having a significant impact on agriculture, with shifts in rainfall patterns, rising temperatures, and extreme weather events affecting agricultural productivity and food security. However, by implementing adaptation strategies, it is possible to reduce the impact of climate change on agriculture and ensure that food security is maintained. It is essential to invest in research and development, early warning systems, and sustainable agricultural practices to ensure that agriculture can continue to play its vital role in the global economy.

Chapter 49: The Effects of Climate Change on Public Health

Climate change is causing major disruptions in the environment, leading to various health impacts. The effects of climate change are not just limited to the environment, they are affecting human health as well. Climate change has been linked to increased frequency and severity of natural disasters, spread of infectious diseases, air pollution, food and water insecurity, and heat-related illnesses. In this chapter, we will discuss in detail the effects of climate change on public health.

1. Extreme Heat

Climate change is leading to an increase in the frequency and intensity of heat waves. Extreme heat can cause dehydration, heat stroke, and even death. People who are elderly, young, or have pre-existing health conditions are particularly vulnerable to the effects of extreme heat. In addition, rising temperatures can worsen air quality, leading to respiratory problems.

1. Air Pollution

Climate change is causing an increase in air pollution levels due to higher levels of ozone, dust, and other air pollutants. This increase in air pollution is linked to an increase in respiratory problems such as asthma, bronchitis, and chronic obstructive pulmonary disease (COPD).

1. Spread of Infectious Diseases

Climate change is causing a shift in the geographic range of many infectious diseases. Warmer temperatures are allowing disease-carrying insects such as mosquitoes and ticks to expand their range. This has led to an increase in the spread of diseases such as Lyme disease, West Nile virus, and dengue fever.

1. Food and Water Insecurity

Climate change is leading to an increase in the frequency and severity of extreme weather events such as floods and droughts. These events can have a significant impact on food and water security. Floods can contaminate water supplies, leading to the spread of waterborne diseases. Droughts can lead to crop failures, causing food shortages and malnutrition.

1. Mental Health

Climate change is also having an impact on mental health. Natural disasters, displacement, and other climate-related events can cause stress, anxiety, and depression. In addition, climate change can have a negative impact on community cohesion, leading to social isolation and a breakdown in social support systems.

1. Vector-Borne Diseases

Climate change is increasing the risk of vector-borne diseases, which are diseases transmitted by insects such as mosquitoes and ticks. As temperatures rise, these insects are able to expand their range into new areas, bringing diseases such as malaria, dengue fever, and West Nile virus with them.

1. Extreme Weather Events

Climate change is causing an increase in the frequency and severity of extreme weather events such as hurricanes, floods, and wildfires. These events can cause physical injuries and death, as well as lead to displacement, food and water insecurity, and mental health problems.

1. Allergies

Climate change is also causing an increase in the prevalence of allergies. As temperatures rise, plants are able to produce more pollen, leading to an increase in the number of people suffering from allergies such as hay fever.

Conclusion

In conclusion, climate change is having a significant impact on public health. The effects of climate change are not just limited to the environment, but are also affecting human health. It is essential that we take action to reduce greenhouse gas emissions and mitigate the effects of climate change in order to protect public health.

Chapter 50: The Effects of Climate Change on Coastal Cities

Coastal cities are home to more than half of the world's population, and they are increasingly at risk due to the impacts of climate change. Rising sea levels, storm surges, and coastal flooding are just some of the consequences of climate change that are already affecting coastal cities. In this chapter, we will explore the effects of climate change on coastal cities and the steps being taken to mitigate these effects.

Sea Level Rise

One of the most significant effects of climate change on coastal cities is sea level rise. As temperatures continue to rise, glaciers and ice sheets are melting, causing sea levels to rise. According to the Intergovernmental Panel on Climate Change (IPCC), sea levels are projected to rise by up to 1 meter by the end of the century.

Rising sea levels have several impacts on coastal cities. First, they increase the risk of coastal flooding and storm surges. As sea levels rise, even a small storm can cause flooding that was previously considered a once-in-a-century event. Second, rising sea levels can also cause saltwater intrusion into groundwater, which can contaminate drinking water supplies and damage crops. Finally, rising sea levels can also cause erosion of beaches and other coastal ecosystems, which can impact tourism and recreation.

Storm Surges

Storm surges are another effect of climate change on coastal cities. A storm surge occurs when a large storm pushes water inland, causing flooding and damage to homes and infrastructure. As sea levels rise, storm surges are becoming more frequent and more severe. In addition, climate change is causing more extreme weather events, such as hurricanes and typhoons, which can cause even more severe storm surges.

Coastal Flooding

Coastal flooding is also becoming more common and severe as a result of climate change. As sea levels rise, even minor flooding events can cause significant damage to homes and infrastructure. In addition, climate change is causing more extreme weather events, such as heavy rainfall, which can also cause coastal flooding.

Coastal Infrastructure

Coastal infrastructure, such as seawalls and breakwaters, is being used to protect coastal cities from the effects of climate change. However, these structures can be expensive and may not be effective in the long term. In addition, they can have negative impacts on coastal ecosystems, such as coral reefs.

Green Infrastructure

Green infrastructure, such as wetlands and mangroves, can also be used to protect coastal cities from the effects of climate change. These ecosystems provide natural barriers that can absorb storm surges and reduce the risk of coastal flooding. In addition, they can provide habitat for wildlife and support fisheries.

Adaptation Strategies

To address the impacts of climate change on coastal cities, several adaptation strategies are being used. These strategies include:

1. Building codes and zoning regulations that require new buildings to be built at higher elevations and further inland.
2. Flood insurance programs that provide financial protection to homeowners and businesses that are at risk of flooding.
3. Early warning systems that can alert residents to the risk of flooding and other extreme weather events.
4. Coastal retreat, which involves moving homes and other infrastructure further inland to reduce the risk of flooding.
5. Restoration of coastal ecosystems, such as wetlands and mangroves, to provide natural barriers against storm surges and reduce the risk of coastal flooding.

Conclusion

In conclusion, the effects of climate change on coastal cities are becoming more severe and frequent, and require immediate action to mitigate their impacts. Rising sea levels, storm surges, and coastal flooding are already causing damage to homes, infrastructure, and ecosystems. Coastal cities must adopt adaptation strategies such as building codes, flood insurance, early warning systems, and restoration of coastal ecosystems to protect against the effects of climate change. It is crucial to take action now to ensure the safety and resilience of coastal communities and to protect the natural beauty and resources of our coastlines for future generations.

Recommendations

Thank you for reading! I hope you found this book informative and enjoyable. If you're interested in learning more about environmental science and related topics, here are five recommended books:

1. "**The Sixth Extinction: An Unnatural History**" by Elizabeth Kolbert
2. "**Drawdown: The Most Comprehensive Plan Ever Proposed to Reverse Global Warming**" by Paul Hawken
3. "**Silent Spring**" by Rachel Carson
4. "**The Ecology of Commerce**" by Paul Hawken
5. "**Cradle to Cradle: Remaking the Way We Make Things**" by William McDonough and Michael Braungart

These books cover a range of topics related to environmental science, including climate change, biodiversity, sustainability, and the impact of human activity on the planet. Happy reading!

Don't miss out!

Visit the website below and you can sign up to receive emails whenever Kenneth Caraballo publishes a new book. There's no charge and no obligation.

https://books2read.com/r/B-A-FXMW-WHBJC

BOOKS 2 READ

Connecting independent readers to independent writers.

Did you love *The Fragile Balance: Understanding Earth and Environmental Science*? Then you should read *The Web of Life: Understanding Ecology and Our Place in It*[1] by Kenneth Caraballo!

The Web of Life is a comprehensive guide to the world of ecology. This book delves into the intricate relationships that exist between all living things, from the tiniest microbes to the largest animals, and the impact that humans have on these relationships.

The book begins by exploring the basics of ecology, including the different levels of organization within

1. https://books2read.com/u/3GVXVK

2. https://books2read.com/u/3GVXVK

ecosystems, the flow of energy and nutrients, and the complex web of interactions between species. From there, the book examines the major biomes of the world, including deserts, grasslands, forests, and oceans, and the unique challenges and opportunities each presents for life.

Throughout the book, readers will learn about the various ways in which humans impact ecosystems, from pollution and habitat destruction to climate change and overfishing. The book also offers practical advice for individuals looking to make a positive impact on the environment, from reducing their carbon footprint to supporting conservation efforts.

Written in an accessible and engaging style, The Web of Life is an essential guide for anyone looking to deepen their understanding of ecology and the intricate connections that make up the natural world.

Also by Kenneth Caraballo

Dropshipping For Newbies
Entrepreneur: Ultimate Guide
Investing For Newbies
The Mind Of A Millionaire: 25 KEYS
Stop Overthinking, Start Living: Simple Strategies for a
Peaceful Mind
Feeling Good: The New Mood Therapy
Financial Stability: A Guide to Achieving Short and
Long-Term Money Management
Unleashing the Power of Brand Magic: A Guide to Transform
Your Marketing Strategy
Focused: Navigating Life with ADHD
Mind Over Mood: A Practical Guide to Cognitive Behavioral
Therapy
The Lost Heirloom
Love at First Sight
Investing Made Simple: A Beginner's Guide to Building
Wealth
Manic Moods: Navigating the World of Bipolar Disorder
From Idea to Launch: A Step-by-Step Guide to Starting Your
Own Business
The Mind-Body Puzzle: Exploring the Philosophy of Mind
The Search for Truth: Exploring the Philosophy of Science

Quantum Worlds: Exploring the Mysteries of the Multiverse
The Secrets of Life: A Journey Through Biology
Inside the Human Body: An Exploration of Anatomy and Physiology
Quantum Echoes
Mastering the Art of Success: Lessons from the World's Most Successful People
The Molecules of Life: An Exploration of Biochemistry
Cosmic Odyssey: A Journey Through the Wonders of Astronomy
ChatGPT: The Revolutionary Language Model that Changed the World
The Wonders of Chemistry: A Student's Guide to Understanding the World Around Us
Breaking the Chains: Overcoming PTSD and Finding Healing
The Voices Within: A Journey of Living with Schizophrenia
The Web of Life: Understanding Ecology and Our Place in It
The Blockchain Revolution: How Bitcoin and Cryptocurrencies are Changing the World
The Brain Unlocked: A Journey Through the Mysteries of Neuroscience
Beyond the Horizon: Exploring the Mysteries of Space
Beyond the Quantum World: Exploring the Frontiers of Physics
The Future of Technology: Engineering the Next Generation
The Beginner's Guide to Being a Christian: A Step-by-Step Journey to Finding Faith and Building a Strong Relationship with God
From Genesis to Revelation: A 66 Chapter Christian Bible Study

The Fragile Balance: Understanding Earth and Environmental Science
The Social Mind: Exploring the Intersection of Psychology and Social Sciences

www.ingramcontent.com/pod-product-compliance
Lightning Source LLC
Chambersburg PA
CBHW070638220526
45466CB00001B/214